Life As

A

Sound Recordist

1. START PROJECT

2. ENTRY CODE

3. SELECT SOUND PATH

4. MIXING IT

5. MAXIMUM GEAR

6. FINE BALANCING

7. NO UNDO BUTTON

8. SHORTCUTS

9. WARNING ARE YOU SURE

10. MASTERING

APPENDIX 1 Connectors

APPENDIX 2 Glossary

APPENDIX 3 Recce

1

START PROJECT

Consider A-levels. More accurately - post A-levels. I thought about becoming an architect but, as it turned out, I had zero aptitude for technical drawing, surveys, ground plans, loading calculations nor any *actual* interest in designing buildings.

Or architecture to be frank, but at school I was interested in both arts and sciences. That was the problem.

Come on, who knows what they want to do with their lives in their late teenage years?

So after masterfully buggering up my exams I really should have had more of a clue, but nope.

Yet from about 12 years old I had been recording music off the radio (illegally as it turns out) on my dad's cassette machine - trying to hit pause before the bleedin' DJ spoilt things, and in my early teens my best Christmas presents were a reel-to-reel tape machine and a guitar. Oh and a bike but that's irrelevant.

Eventually I built my own stereo system, speakers and sound-to-light display in our bedroom with the reluctant (i.e. enforced) help of my younger brother. I got into sound-on-sound, track bouncing, early cheap synthesisers (such as the Wasp) making it all up musically and technically as I went along but I didn't know what I was going to do with my life because none of this stuff was in the "Book of Jobs".

You would think, in hindsight, that it was obvious, but it honestly wasn't to me at the time.

Early days - late nights. I was blissfully wearing headphones, listening to some pretentious prog-rock, unaware the main amps were still on. My parents came home to find the house thumping to Emerson Lake and Palmer. I believe they had to apologise the next day to our poor, suffering neighbour.

He put up with a lot did Mr Brown. I'd like to say sorry too, but it's a bit late. I could apologise to my brother but... nah.

Sound and music were important to me yet...

it surely couldn't exist as an actual job could it?

That was for famous people who produced their songs and sounds by magic.

In my defence there was no available internet, and the local library did indeed have a "Book of Jobs" but there was nothing under the heading - "Messes about with Sound".

There wasn't even a heading -"Sound". There probably still isn't. Anyway, who uses books to find things out these days? Oh yeah...

It did have chapters extolling the virtues of becoming an Actuary, Dentist, Market Gardener, Undertaker or Zoologist, or, if you were of the female persuasion, Receptionist, Secretary, or more ambitiously still, a Personal Assistant (attractive ladies only need apply)

Yes - the book was that old-fashioned, sexist, middle-class and entirely useless, cobbled-together by an old-fashioned, sexist, middle-class and entirely useless committee no doubt.

School-based job-advice was hopeless too (information and inspiration-wise) Maybe it's better these days? I hope and believe things are improving and I am sure there are many excellent teachers out there who are concerned for your prospects and future welfare.

I just never met any.

Did they ask: "What really interests you? What are you good at?" and do a little research, or did they give you some leaflets to throw away? It's not my place to criticise the education system of this country. In many ways it is excellent.

In many other ways, it is not.

But never mind. Here, at last, is a "sound job" guide.

After giving up on further education, eschewing University and through a series of both chance and misfortune (like getting sacked from a job in accountancy in an advertising company that I hated anyway), I somehow ended up at Television Centre, London.
Apparently, at the time, it was "a cultural icon... a factory for television" and a "powerhouse of creative broadcasting."

I wouldn't disagree.

What mattered was that in those Halcyon days, the BBC was a huge and magnificent drama and entertainment machine, and its hub was TVC - Television Centre, White City.

The Emerald City it wasn't - well not quite - it was a lot more popular.

Through luck and circumstance I had ended up where I needed to be without knowing it.

My temporary summer job at the time was totally unrelated to sound, but I was at last in an environment where I began to come across rather intriguing and previously unheard terms such as –

boom operator, grip, sound supervisor, studio manager, focus puller, grams operator...

None of these professions was in the bloody "Book of Jobs"! I have no idea why not. I suspect the publishers were incredibly lazy and didn't care, or perhaps it was because I had discovered a rather secret world – unknown outside its own domain, yet full of promise.

I decided to find out more.

Fortunately, the BBC at the time had a policy of posting all available internal jobs on the walls on every floor of that circular building and those cork-boards became irresistible reading:-

> *Available posts –*
> *Audio assistant, Trainee camera operator, Floor manager, Senior sound supervisor, Costume movement operative, Engineering manager, Catering assistant, Security commissionaire*

- all were actual jobs in one glorious mix! They really did advertise for "Head of Programming BBC1" alongside "Junior Payroll Clerk", and these posts were updated every two weeks, with detailed requirements for what was expected of the candidate.
It was extraordinary and yet it seemed to me the majority of the denizens of TVC walked blithely past these opportunities every day, content in their own spheres. Which they probably were.

Of course I knew nothing.

Messing about in your bedroom with a reel-to-reel is not actually that impressive to a BBC interview board of five stern-faced professionals as I found out.
Yet I was able to watch productions in TVC studios One and Two from the viewing galleries, occasionally sneaking onto the studio floor during lunch breaks, and I took advantage of

joining "BBC Club" amateur organisations that (amazingly) allowed some of us to play with real BBC radio equipment in the evenings (as long as we passed a basic competence course and faithfully promised not to break anything - including ourselves)

A group of us were allowed into a radio studio one evening to produce some amateur radio play. We were quite excited because this was a really shiny new studio – not the creaky ancient basement facilities the BBC usually consigned us to.

Everything would have been fine if us cocky would-be studio managers had known where the ON switch was. Turned out it wasn't on the desk as expected. Took us twenty minutes to find it. And that was only because someone else (totally non- technical) said: "Is it maybe this big red switch by the door?" Remember that Mike?

It has been a while (to say the least) since I started my sound career but I see no real difference then to now – you need dedication, persistence and perseverance if you are going to lever your way into this business

 - in other words you need to be pretty bloody stubborn and competitive.

And wildly talented, which helps.
Also full of enthusiasm, energy, and a "joie du son", if there is such a thing.

It's a lot easier now to find out about these things and put in the research, for example via Google, Bing, Facebook, or Yandex (please note other search engines are available) but is it any easier to break into the area you are interested in?

It's probably harder, actually, but don't let that put you off!

What do I know? Except...

It does help if you can display a bit of talent (though that is by no means a pre-requisite as I know some who have done very well on charm alone). But that's the same in almost every walk of life, isn't it?

Still, it is difficult to totally bluff your way through the film and TV industry because you will often be surrounded by experts – people who not only know their jobs, but know yours as well. They will suss you out in 30 seconds. Look on that as a positive thing. If you actually know what you are doing then, in all likelihood, your colleagues will see that and give you a nice tick/check mark on their private contact list.

And you never know where that might go.

Sometimes it even leads to that one phone call, email or social media contact that changes everything.

Jobs in the freelance world depend upon word-of-mouth, recommendations, the quality of your work, personality and a shitload of luck.

I got a call out of the blue late one morning from a production co-ordinator practically begging me to work in their studio that afternoon and evening. Well, it was a pain and it wrecked my plans for the day but I went in as asked, and helped out on a new show. It was called Kings of Comedy, and nobody remembers that now, but that one act led, I kid you not, to years of work with that company, because they provided the crew for Endemol - just as *Big Brother*, its spinoffs, and *Deal Or No Deal* exploded onto the scene.

Once it sniffs success a business will want to follow that up and TV production companies are more tenacious and committed than most because they know their product has a limited lifespan.

If you can spot that opportunity and jump on that bandwagon, you could be living that dream.

Well, pretty close anyway.

2

ENTRY CODE

I have had the benefit of BBC training, though I had to work for it. BBC Woodnorton was rather like Hogwarts without the fun magicky bit. It even came with an irascible warden called Ernie.

They took you right back to the fundamentals (we actually had to learn how to do equations again) but slowly we progressed to practical applications, and it became more enjoyable, except for the weekly tests. Fail three and they chucked you out - back to loading pantechnicons, checking expenses forms, making tea or whatever you used to do.

Bogwarts aka Woodnorton Hall
Courtesy Liz Horne

It was good motivation, I'll give them that.

Bit by bit, hardly any wands or potions involved, we learnt the magic of television.

We made films and videos and played at all the jobs in a television studio. We made a radio series - "The Wrath of Zog" that has strangely disappeared. We were shown the stairs down to the nuclear bunker where the BBC would hunker down in the event of WWIII. We brought in our guitars and drums and recorded a terrible racket. We were hunted by Ernie for letting off the fire extinguishers. And a few weeks or so into the initial 3 month A-course some of us discovered that getting plastered in the Phoenix Bar the night before a test actually improved our results. Well, that's what we chose to believe, and it was certainly more lively and sociable than studying all alone in your room. I think it relieved stress. It certainly seemed to work. Of course, some people weren't quite up to it. I had never before seen anyone *actually* turn green prior to throwing up immediately before a test but Timmy managed it rather spectacularly (and noisily in the bushes).

I never appreciated at the time how lucky we were. Well, you don't do you? But none of us got a single piece of paper saying: "Trust me. Despite appearances, I know what I am doing."

The vast majority of A95 are still in the sound business.

There are no sound qualifications as such in the real world, though I have no doubt the BBC gave me the best possible preparation. But I have no certificate, no diploma to prove what I know. I'm just thankful for the guy who took me through those frequency/wavelength
calculations before that sphincter-clenching interview when I got accepted into an Audio Unit (cheers Geoff)

Do you need to make those calculations as part of the job? Of course not. It's just groundwork – like O-levels. You don't need Sabine's formula to work out whether an acoustic is crap or not, but the BBC thought it might be useful, so you were going to learn it, or else.

As a course, we decided to go to a Kinks gig one night in nearby Birmingham. Not perhaps the best decision in the world as we had a "stereo appreciation" class the next day and all any of us could hear was a sort of muffled buzz. It turned out OK though because the emphasis was on low frequency dispersion.

Lecturer: Can you hear that low note?

    Yes thanks.

Where is it coming from?

    That speaker over there.

You're not supposed to be able to spot that due to the low frequency.

    Really? Frankly that's all we can hear.

The motley A95 crew. So young. So thin. So hairy

After proving you can be trusted to work for the BBC without electrocuting anyone or editing your own fingers off you get sent to do Qualifying courses and Studio Management courses and Managing People courses, all of which are rather jolly so long as you maintain perspective and the right attitude – I'll expand on this later, if I remember.

As part of my training I also had an attachment to BBC Birmingham where I learnt to edit *The Archers*, boom op on *Angels*, and work with Kiki Dee and Leo Sayer. Well, it was pretty exciting for me at the time. I was young and had never met anyone famous at all, so gimme a break. Leo's mic flew off when he was swirling it around, as he frequently did.
That might just have been my fault.

After training you go back to base (in my case, the splendid Bristol Audio Unit) and the real fun starts.

I've worked in a wide range of sound activities – television, radio, studio, outside broadcast, film, video, classical and contemporary music, drama, documentary, natural history, live broadcast, and post production. I am an expert in none, proficient in some, but at least I can say my experience is comprehensive.

In the days of quarter-inch[1] tape there once was a green audio assistant who cheerfully and skilfully edited a radio program in less than an hour and, rather pleased with himself, looked forward to the rest of the day off. He then proceeded to make a backup copy of the tape as was normal practise. Unfortunately he was distracted and flirting with the attractive producer and pushed "play" on the record machine and "record" on the playback machine with the edited program on it. He spent the rest of the day trying to repair the mistake and the rapport. Unsurprisingly, he never got to go out with the producer either.

Perhaps not as bad, though, as a colleague who cut through a whole ten inch [NAB][1] reel of tape with a razor blade believing it to be gash, only to discover to her horror that it was the actual whole programme. I believe it took two days to re-assemble the two thousand eight-inch bits.
Nice work eh Sue?

Editing is a *very* useful skill, and having some kind of inkling as to what will cut and what won't is incredibly helpful. The methods might have changed but the principles will always remain the same. That said, though most people can learn how to edit digital media fairly easily, there is yet an indefinable sense of timing that separates the competent from the seriously talented.

I have also served my time rigging microphones for chamber ensembles, orchestras, bands, football and rugby Cup Finals; I have built temporary OB[1] studios from the ground up for live productions – wildlife, game shows, entertainment and news – and mixed those shows.

I spent many, many years in studios and on location using booms[1] of one sort or another, and learning the whole sound "trade", and getting into the after-hours social life. It was fun, and we got paid to do it.

Best of all was working on location either in your own country or somewhere more exotic for days, weeks, or even months.

Getting to work and bond with a crew is one of those unseen bonuses. It's important, and can lead to lifelong friendships.

There is a downside, however, and I'll address this later.

[1] Like many professions, the film and broadcast world is full of jargon. Please see appendix 2 for a translation.

A typical early OB lash-up unfortunately 278K from the nearest pub.

Blimey Rod, what a mess.

Live shows, Outside Broadcasts, music recording, wildlife, news – it's all great fun.

However, a very great deal of my working life has been spent as a freelance drama sound recordist/ production sound mixer and that is what this guide is about, with a few extras thrown in for good measure and padding.

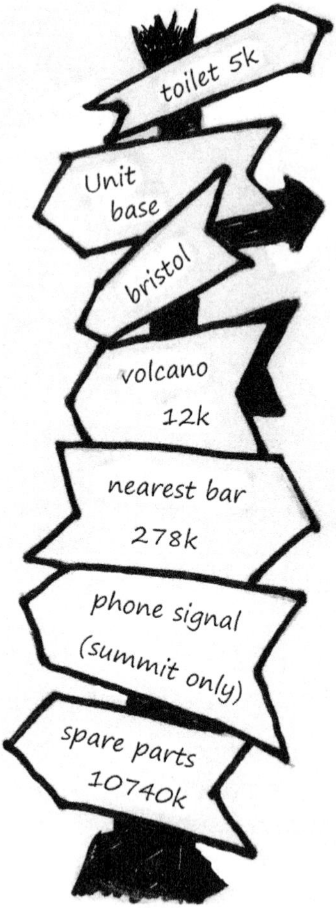
toilet 5k
Unit
base
bristol
volcano
12k
nearest bar
278k
phone signal
(summit only)
spare parts
10740k

3

SELECT SOUND PATH

WHAT is a Sound Recordist anyway?

(Note that, especially in America, recordists are known as sound mixers, or production sound mixers. I prefer recordist because it's a more honest description, but production sound mixer sounds a bit grander and more important doesn't it? And we all want to be important.)

When awake – he or she records sound. On a sound recording machine. Easy.

What we do:
Either alone or (more usually) as part of a team you set off with your equipment to some possibly distant location to record some sound there. What and how you record depends on the remit.

What we aim for:
Simply put, it's to get the most wanted sound in comparison to the unwanted sound – it's called the signal to noise ratio. A couple of decades ago that meant something more technical (in the days of tape hiss) but forget that - it's all about hearing what you need to hear against all that background crap that the world is trying to fling at you, and the world is trying *really* hard.

What you want to hear (and record) is usually a voice, though it might be a violin or a cricket, a lion or a volcano. You get the idea – it could be anything audible (or even inaudible, in the case of bats or some particularly mumbly bloody actors from the school of internalisation)

We aim to separate the wanted from the unwanted. You might think, for example, that an eruption is frikkin' loud and therefore dead easy to record, but you would be wrong.

Crickets (*if* you can get close enough without them mysteriously going quiet) are really strident. So are violins, though it depends on how and where they are being played (and by whom.)

In every case, what is happening in the background is important, and it depends how close you can get to the source of the sound you actually want that will largely determine the quality of the recording - though not the quality of the playing, of course.

Is your violinist in a soundproof studio, or in the open in Covent Garden, or in your front room? Is the cricket in the middle of the South African veldt or on the local common (next door to a primary school at lunchtime?) And when the volcano spectacularly explodes are you on your own, prepared and ready with remote microphones (having waited for endless weeks) or (more likely) embroiled with a bunch of panicky screaming bloody tourists because you just happen to have been there when the thing went bang?

Of course if you are a videographer that doesn't matter so much because the sound matches the images but the sound recordist is aiming for the pure and undiluted because that is what we do. At least, that is what we hope to do. It doesn't always work.

Arguably the easiest of the above examples is the violin as it is highly likely it will be in a controlled environment – but the director may well have other ideas, such as recording the performance on the top of a cathedral in a storm, say, because it's a more dramatic visual

interpretation of the piece.
Well, it might be dramatic but it's a pain in the bloody arse. And inadvisable health and safety wise.

Also, crickets, volcanos and most of nature will generally refuse to co-operate... until the very moment you decide to pack-up and return, in defeat, to the hotel. But then... wait a minute... was that a Panamanian Night Monkey?
Do you unpack all your gear again? It sometimes depends on the hotel bar, how comprehensive it is, and the generosity of the producer.

Actually, a lot of it depends on that.

If it were that easy, any stupid bug... could do it

What exactly is a good sound recording?

• It has a high signal to noise ratio i.e. it has a lot of what is wanted and very little extraneous garbage.
• It is not over or under modulated (clipping is verboten; under recording is technically noisy)
• It has a dry acoustic or an acoustic appropriate to the scene (no unwanted echo)
• It fulfils the remit (if asked for a lion roaring it's really no good giving the client an excellent example of a lion farting just because that's all the bastard animal did all day)

Of course you can guess what I am going to add here - sound recording isn't just about making a good sound recording. More on that later.

Everyone has their own ideas.

Chris Watson - The Art of Location Recording:

Over the years, Chris has had time to reflect on the quality and quantity of his work, and nowadays chooses to record far less material than he once did. "I am very careful about pressing Record now, because I used to record everything, and realised that a lot of what I'd recorded wasn't very good. That has made me more careful and honed my creative process. I now listen more than I record. It has to be worthwhile and meaningful for me to make recordings. That connects me to the recordings

in a personal way, whether it's the habitat, landscape, animals, people or the processes."
[source: Sound On Sound]

Steve Morrow (La La Land sound mixer)

"Sound Mixing includes both Post and Production sound. For Production sound mixing, the main goal is to record the actors' performance as cleanly as possible. This is mainly the dialogue of the film or, in a musical, the singing. I also like to gather ambient noise and location-specific sounds that may help the Post Sound team. I always put it like this: if the movie's sound is a gold bar, I'm the strip-miner grabbing as much good material as possible so the guys in Post can find the golden nuggets."

Production Sound Mixer Ronan Hill (Game of Thrones)

"For me, it's never enough just to record the dialogue, and we try to record stereo effects in addition to a mono boom where possible, to get a real sense of drama. We also fitted mics to cameras on tracking crane arms and radio mic'd a few horses, placing the capsule on the girth to get clean horse hoof effects for the charge into battle."

"In addition to boom and stereo mics, Jon Snow would have been radio mic'd with particular

attention to mic capsule placement and transmitter placement. When you fit a radio mic to a person their body absorbs some of the signal. The fact that he is surrounded by a large compressed crowd makes getting a radio signal from his transmitter very difficult, but it is essential to record the integrity of the original performance."

Unsurprisingly, it's not all unmitigated fun. Season 7 required Ronan and team to work in Iceland:

"Most of the locations were remote, travelling in by four-wheel drive and then walking and pushing the rest. The weather was hostile with snow, rain, hail, high winds, and temperatures well below freezing. The first day we had gale force wind in excess of 60 mph and wind chill below -30C." [source: Sound&Picture]

Sooo... Are we having a re-think yet?

A brief rundown on other sound professionals:

Sound Assistant

These indispensable guys and gals work on the studio floor, the set or location, doing all the hundreds of jobs that need doing – setting up microphones, speakers, handling transmitters, rigging artists or contributors, running cables and plugging up, laying mats, fixing faults as they occur, moving equipment as necessary and all as quickly and safely as possible without upsetting anyone else. It's where most learn the trade, and where you learn whether sound department is actually for you, because you get very little recognition (to be fair though, that goes for most departments/trades/skills in this industry)

There is a lot to absorb, and common sense, personality and aptitude are key. You get out what you put in. Some assistants are so good that recordists/mixers (if they are honest) never want to see them move on, but move on and develop they must. Some sound assistants are so bad you hope they break early, slope off, and go on to be lawyers or estate agents instead (they'll make more money and never have to run anywhere to get their job done so it's not an unreasonable wish)

Without reservation I salute all Sound Assistants out there, and thank you for putting up with so much.

I also did my time and yes, of course it can be a lot of fun. Hang on to that thought even as you wade through the mud at midnight searching for the van keys you lost because if the Recordist finds out you are so f*cking dead. Did you try the fridge? Of course you did. How about that secret waterproof pocket? And oh shit what about that pile of cables left behind the barriers on the road to nowhere? You know who you are.

I have now put a tracking device on my keys. Next step is to put a tracking device on the assistant - now that's a bloody good idea!

Boom Operator
This is the person who holds a long stick with a microphone at one end. A skilled operator is probably one of the most undervalued people on set because, frankly, they make it look too easy. Their aim is to get the microphone as close to the relevant artist as possible without causing any problems (such as appearing in shot or causing a shadow) whilst looking cool and unperturbed, and drinking coffee.

Where the F*** did that come from? We're in the middle of bloody nowhere!

If this section sounds like a bit of a love-in then I don't care:- you try running backwards in a ploughed field with a big weight on the end of a 3m pole ten times in a row without cursing, falling over or injuring anyone else.

Oh yes, and the microphone must be pointed in the right direction all the way and there mustn't be any audible bumps and the furry dog cannot bounce into shot.

They have eyes in the backs of their heads (situational awareness); a high level of fitness; an ability to communicate with artists, the director, camera team, DOP and you. They are not necessarily big and strong but have poise, balance, stamina and a lightness of touch.

A good boom op is worth their weight in decibels. But for God's sake don't tell them - they'll want more money.

Boom ops usually carry around a wooden box that they can use to stand on.

It has holes in for the boom poles, but more importantly space inside for their coffee machine, gloves, antiseptic wipes, gaffer tape, hairbrush, makeup, powerbar snack, water bottle, banana, shorts, deodorant, sunglasses, baseball cap, chewing gum, ninja clothing, spare socks, phone charger and Life As A Sound Recordist paperback.

This convenient boom-holder is designed to look like a real person

Sound Engineer

Generally speaking someone who works in a mixing suite at the back doing a lot of hard work for little thanks, probably in a music studio or a theatre, but the term covers a whole multitude of jobs.

WIKI says (and I mostly agree, for a change) -

"An audio/sound/recording engineer helps to produce a recording or live performance, balancing and adjusting sound sources using equalisation and audio effects; and in the mixing, reproduction and reinforcement of sound. They are involved in the technical aspect of recording – the placing of microphones and the setting of levels. The physical recording of any project is done by a sound engineer. They also set up, sound check and do live sound mixing using a mixing console and a sound reinforcement system for music, concerts, theatre, sports games and corporate events."

The Sound Engineer is not to be confused with a sound maintenance engineer who works alongside a recordist in the field as an assistant or a boom op, pretending to maintain the kit for extremely boring tax purposes.

Sound Supervisor/Mixer

There is a lot of nomenclature crossover here depending on your country of origin, but the sound supervisor is generally the studio or OB-based top banana. They are not usually part of the management structure (although they can be in smaller firms) but they are always highly experienced, extremely valuable, and well paid. They use desks the size of a tennis court to mix "Saturday Night Dancing Fever on Ice LIVE" or whatever. They are experts at pressing buttons and achieving mixes from a hundred sources, sending multiple monitoring mixes, and matrixing everything in such a complicated way that nobody knows what the frig is plugged to where so they had better not get sick (although their number 2 probably knows more about what is routed to where than they do to be honest, and is secretly hoping the boss *does* get sick for promotion purposes)

Still, it is an incredibly involving, rewarding and skilled job. You must enjoy being in a dark room for most of the day and playing with lots and lots of buttons and flashy lights. You must want to sit back and look at the sound desk like it is your fiefdom. You must want to be asked to sweeten the first violin or fatten the audience because you know exactly which knob to adjust. And you are allowed to have it all at deafening volume with no neighbours to complain.

It's about the Power!

When you finally accept that films and television are nothing without sound, you are prepared to defend that position to the hilt.

What is a dance show without music? Does anyone want to go back to the days when a lone organist accompanied the film?

Didn't think so.

It's like being Jafar the Genie.*

"Sound is not understood by very many people," warns Tony Revell, Sound Supervisor, whose credits include Strictly, the NTA's and BAFTA's.

"Producers, directors or anyone on the visual side of television really don't know what sound involves or how it is created."

Building a career in sound, Tony believes, is the same as it was when he began over 40 years ago. "[You] need to do the work at the coal face, and work your way up from Sound Assistant to Sound Supervisor. A degree isn't always necessary either. Due to how specialised TV sound is – the skills needed differ greatly between genres - there aren't many specific courses.

I don't think I know who has done a degree. It's never come up."
"You have to interpret the production's requests in a way which is feasible to achieve, by knowing what is or is not possible, and be able to explain

that to the rest of the production. You have to develop cunning ways of getting what you need without directors shouting at you because... the microphone is in the way."

Sound, he says, is as much about cutting out what you don't need as it is capturing what you do.

"It conveys the majority of the programme... [so] if something goes wrong then people start shouting." As such planning for when things go wrong is essential. "If you can actually hear something is wrong and go about fixing it, you are 90% of the way there."

[source Tony Revell]

* I did mean Jafar The Genie - Jafar is a Sorceror who wants ultimate power. Check out Disney's Aladdin (1992 version). Awesome sound mix.

Post Production Sound

This is a whole different enchilada that includes a panoply of editors, re-recording mixers, foley artists and so on. In fact there are so many different jobs in sound post production (especially if we are talking big feature films) it is almost impossible to list them all. Suffice to say they generally work in a dark, soundproofed cave and rarely see the light of day, rather like the Panamanian Night Monkey, unless they are released to record some much-needed real-life munitions and machine guns, or exotic cars for example, a job anyone with an interest in sound should bite their own leg off to do.

Post rarely emerge on set or on location. This is usually because -

> (1) the real world is a strange and uncontrollable thing
>
> (2) the sound recordist will grab them and explain in great detail why the sound recording is so difficult on this particular shoot and
>
> (3) the DOP will grab them and explain in great detail why filming is so difficult on this particular shoot and
>
> (4) the Director will grab them and explain in great detail why everything is so difficult on this particular shoot and how it must be fixed in post.

Re-Recording Mixers

Re-Recording Mixers, otherwise known as Dubbing Mixers, work with all the sound elements (dialogue, automated dialogue replacement, foley, sound effects, atmospheres, and music), and balance them to create the finished soundtrack for a project. They are primarily responsible for ensuring that the final sound mix is correct both technically and stylistically. Setting the relative volume levels and positioning these sounds is an art form in its own right, requiring a high degree of skill and aesthetic judgment.

They spend a lot of time in moodily lit rooms full of eerily glowing equipment, posh seating and beautifully polished speaker cabinets and the occasional extrusion of rustic wall. This is the job for you if you enjoy polishing, cleaning, buffing, balancing and stamping your own soundbadge on a project in conjunction with the director and producers.

It can be highly demanding, and extremely rewarding.

Because of changes in technology, many jobs in sound post production are less easily defined, e.g., on some small to medium budget films, Re-Recording Mixers may also work as sound designers. Tech marches inexorably on and now post production has to cope with 2.1, 5.1, 5.1.2, 9.1.4 ,Dolby Atmos, DTS:X, FFmpeg and so on.

The industry wants (and needs) the public to constantly update expectations and their home cinema, so, alas, far from getting simpler, sound has become even more complex – presenting its own challenges and opportunities.

I don't have a problem with this – although I have to say a pair of good £200 - £300 headphones is all you *really* need most of the time. Yet it is a wonderful thing to have a full setup at home – you might have a wallful of 8k video but it is nothing if not accompanied by the complementary quality sound. Perhaps I'm shallow, but I love my B&W setup for the music and the explosions.

You can't hear bangs in space – it's impossible. So why do we get to hear so many cosmic explosions and the like?

I know of a young, keen re-recording mixer who confessed to me that while dubbing a natural history film about whales, and before the grown-ups arrived to review his efforts, he decided to make it all more realistic by changing the timing of all the distant whale splashes as they breached - perfectly logical - we all know that sound from half a mile away or more takes a bit of time to reach our ears. Of course, the director and producer were aghast - why is it all out of sync, they demanded?

"But it's like real life" he protested.

"Change it back, now" they said.

Good try JP, but doomed

Pace, tension and emotion are all deeply affected by the accompanying sound. We don't always *want* realistic.

Most Natural History sound has been cobbled together by foley artists over the years (though not all, I must point out)

Pull a sword out of a leather scabbard and it does not go "Thhhhrrriiingggg!" I regret to say.

Suppressed pistols do not go "Phhhtt."

Laser beams are silent.

If this is a surprise to you then get ready for a lot more - ants do not make audible scritchy-scratchy sounds and I know the name of the guy who did the baby polar bear impression in that famous shot of mother and pup emerging from their snow cave.

Sorry.

The following job brings all of the sound elements together - from conception to invention; recording and mixing, and overall control.

I wish it had been in the "Book of Jobs", but it didn't even exist back then.

At least I can tell you something about it now:

Sound Designer

The Sound Designer works closely with the Director and a range of other staff to create the aural world of the show in a post production broadcast or theatre environment.

The Sound Designer can create sound effects, atmospheres, sonic textures and filmic ambiances that will fit the worlds of the project's story, as well as aid the audience's emotional and dramatic connection with the performance.

The Sound Designer may choose, edit and remix recordings; work with a composer to make original music; or work with live musicians in the theatre.

The Theatre Sound Designer may advise on how to best hear the performers, which may involve acoustic adjustments to the set, or the addition and configuration of mics for the performers and might even involve design of a sound system, bespoke to the specific production and auditorium, that will give the audience the best experience of the show.

The Film Sound Designer could be said to be the "author of a soundtrack"

Gary Rystrom, asked about the title of Sound Designer

"Well, for the Academy Awards, for the sake of getting nominated, I'm technically a supervising

sound editor. But I take the extra, highfalutin' title of sound designer for almost historical reasons. Because where I work, in northern California -- the same as Ben Burtt did sound design for the "Star Wars" movies, and Walter Murch did sound design for "Apocalypse Now" -- there's a tradition of people I know who are mentors to me. In the area of the country where I work, there's a history to that term, so I come at it historically. What I like about it is that it implies someone who really is the author of a soundtrack, so from the very beginning -- the first discussions with the director -- to the end of the mix, I feel like it's my job to design the soundtrack. So I like the term, although other people might think it's a little too big for its britches."

Sound design is challenging, fun, stimulating, warm and dry - an interesting field when you have enough of trudging around *in* the field.

I like it, though it has its many challenges. I have had a bit of a play on a very small scale.

WHAT FIELD SHOULD I CHOOSE

You don't always get the chance to work in your chosen field because when that production company contacts you out of the blue regarding an exciting-sounding 3 month project, all your preconceptions and high promises to yourself will immediately go out the window because:

- you probably need the money and

- your most powerful initial feeling is "Oh My God somebody wants me" and

- it's an adventure and a challenge you had never previously considered would come your way.

You should go the way your heart advises. But maybe your head is saying otherwise? Or your wallet.

If you wish, you could make a name for yourself as a sound specialist that gets themselves known for doing one thing exceptionally well, e.g.

> Shorts Corporate Docs
>
> Commercials Reality
>
> TV general TV Drama
>
> Nat Hist Feature Films

I have left a load of stuff out because it kind of comes under another sound remit (studio or outside broadcast for example) There is also a lot of difference between live transmission and recorded material.

I cannot recommend one course as being better than another. If you have set your sights on being a big Feature Film mixer then good luck to you, and I mean that honestly.

There is tremendous fulfilment to be had from Natural History recording, but it's sometimes a lonely affair. Drama is great as long as you are a team player. Commercials are well-paid but repetitive and sound is largely post. Documentaries range from the downright dangerous to the deliciously deluxe. Reality TV is technically challenging, but hardly involves any artistry (that's all in the edit, but perhaps that appeals?) Corporate gigs range from the well-paid and ridiculously easy to the insulting.

These are just my opinions, of course – a distillation of 40 years experience. Others will disagree or have different outlooks; fair enough – please read their books.

In the normal course of events you will offered a variety of jobs over the years, and you will gravitate towards whatever area stimulates you most. One of **the** *worst things* about being a freelance worker is that job opportunities will overlap. You'll then have to decide which you are available for based on the type and length of job, the money, your personal circumstances, and the likelihood (or not) that it will lead to more work.

These decisions are heartbreaking because you will never know what might have happened – but I strongly advise against going down that road.

46

Regretting what you didn't do is as dangerous and confidence-sapping as it gets. It's even worse than getting envious about that cream job that another professional - probably a mate, or someone known to you - manages to land.

> Prefer the sharp end? Want to get cold wet and dirty?
> Rely on your own initiative?
> Enjoy camaraderie under pressure?
>
> There's the Marines, of course, if you like guns and stuff.

Or maybe you want to be a bit more creative?

4

MIXING IT

WHAT YOU NEED to be a SOUND RECORDIST

Ears

- no, you don't need owl or bat-ears, but you do need normal hearing. You can get your hearing checked easily. Human range is roughly 20 to 20,000 Hz. If you don't know what that means, I'm afraid you have some research to do. Younger people have better hearing in general, but that is not the whole story. A sound recordist needs to listen. To actively listen is a skill that can be learnt, and fairly easily too, but you'd be amazed at the number of people who don't listen. Or maybe you already noticed that, in which case please read on...

Something Between the Ears

You really are going to need some kind of technical ability in this game. Unless you have limitless amounts of money and can afford backups of backups beamed up to the cloud in, say, the Atacama desert via your own personal satellite, you will have to be able to fix things (or at least cobble something together) when it all goes tits up.

Which it will.

I had to fix a cable whilst bobbing about in a boat, using its 12V battery and a soldering iron I had fortuitously loaded with my gear. At least it was the Caribbean and not the Atlantic. We were filming sharks at the time, with a presenter doing underwater ptc's whilst the sharks swam about in the background and this simple bit of kit was vital. Motoring back to port would have lost a day's filming. So I advise, learn how to solder... in unexpected circumstances.

You should also know about microphone characteristics, digital recorders, timecode, frame rates, RF theory and electrical safety.
Other useful skills are sewing, first aid, off-road driving, navigating (satnav is not ubiquitous) and other craft skills (basic camera and lighting for instance) If you can squeeze in cooking, fishing, mechanics and simple stringed-instrument or bongo entertainment as well then

boy have you got a career ahead of you!

It's rare to be able to fix modern equipment in the field these days unless it is just cable failure (which it mostly is).

However, identifying technical problems is crucial to keeping the production running and not holding things up. Problems will occur, especially in adverse conditions, and if you can tell where the fault is then you will save yourself a lot of trouble and save the production money (for which they will be grateful and they might even employ you again. Naturally, the opposite is true if you DO hold things up)

If you are a part of a big team you may be able to pass the problem on, but that won't always be the case. You need to be able to improvise. The recorder has packed up so can you record on the camera? It is ALWAYS a good idea to look up the camera you will be working with before the shoot to find out what the audio input is - XLR, Lemo, 3.5 mm, 5-pin? They vary so much it sometimes feels like a conspiracy, but it is more to do with ergonomics.

My pet hate is the 3.5mm connector.
They vary in size (depending on where they were manufactured) so they make intermittent contact. They are horrible, small and cheap, and constantly fail but are used even by reputable manufacturers for, say, headphone outputs.
But they are shit. I learnt a harsh lesson when one of these bastards snapped off inside a laptop

I was using for playback - just the very end bit.
How would you cure that in the field?
Answer in Appendix 1

Oh bugger - digital problems. i.e. fingers pressing the wrong buttons. Otherwise, the pressure is, um, intense.

Sound equipment, properly maintained and protected from the elements is pretty robust, generally speaking, but water is a bastard (let's not even mention salt water.) There is a very distinctive intermittent fizzy/crackling sound when water gets into the system. You'll recognize it. Not too bad on a documentary perhaps (with a quick cable-swap) but on a big drama with multiple single runs on a rainswept night-shoot it's a bugger...

Oh and don't for a second believe that radio mics will save you.

The Atacama desert, by the way, is where you will have the least problems in the world regarding water ingress.

Replacement is often the only cure, but the digital mixer you get flown in will NEVER be set up the way you want so you had better know your way round the operating system and nested menus. An awful lot of technical manuals are available as pdf's accessible online but nothing beats your own edited pencil notebook version stuffed in that secret waterproof pocket for those times when *there is no signal*. It is indeed, a wonderful feeling when the emergency replacement bit-of-kit actually lights up, (oh thank you nameless in-house engineer who actually charged the f*cking batteries) but then the real work begins.

You may have been given equipment with which you are unfamiliar. In a complete emergency situation concentrate on getting one channel working and routed to an input of the record bus. Don't forget about phantom power. Make sure you are monitoring the right thing. Record something and play it back. Forget about metadata and distribution. Is it usable? Ready to go? Good. Download the manual pdf later in your hotel room.

When you are happy, maybe you can join the rest of the team in the bar in the evening and have that very-much-needed drink and, possibly, a cry on a supportive shoulder (depending on the image you wish to project - sensitive, artistic, empathetic, macho, enigmatic, solid, whatever. As a recordist, you have a nice wide choice.)

Certain crafts need to project a complementary image - you don't get wimpy grips or fey sparks. Camera operators are never shy. DOP's are confident, artistic technocrats. Costume Designers are expected to be slightly eccentric and a touch flamboyant. Make-up artists are unfailingly cheery and attractive. Stunt co-ordinators are tanned, ultra-professional and drive huge black SUV's.

About Timecode

This is a complicated area that simply needn't be. I suspect it will eventually be superseded. I am not going to go into great detail here, suffice to say you need to have an understanding of why it is important. If you ask most recordists what 24.97df is for they probably won't be able to explain because it is fiendishly complicated.

There is an excellent guide to timecode here:

https://blog.frame.io/2017/07/17/timecode-and-frame-rates/

I highly recommend it. Basically, everything is so much easier if you live in a PAL-based country like the UK, or in Europe or Australia. But of course, one day you might be invited to work for an American or Canadian firm! Americans will already be familiar with the pitfalls.

To quote from that source - "timecode is native to your files, so bad timecode is basically like a genetic defect."

So, pretty serious then.

When "talkies" were invented there had to be a means of synchronising the sound and the film which were recorded on different devices. It was dead simple and involved a slate or clapperboard. So the editor sees on the slate "scene one take three" on his Steinbeck and marries it up with the sound "scene one take three CLACK!" and amazingly the rest of the scene is in sync. Usually.
Assuming no problems with the camera or sound gear. Or the transfer suite. Or... oh never mind.

And that's how it was for a long time, all held together by sprocket holes and mechanics, until video recording came along and the sound was recorded on the same tape as the vision – so that was all right then, problem solved.

Except that the video quality was execrable, so it was never used on anything big-budget, but a massive amount of stuff was shot on tape formats where sound and vision were locked (though they could be separated in post production of course.) TC helped with synchronisation, scene logging and editing tremendously. It was a very early form of metadata that meant both sound and vision were stamped with a synchronous code that related them and therefore you had no real need for slates.

As digital formats developed, however, so sound and vision went their separate ways again (much to the relief of both departments) but mysteriously and crucially linked by timecode.

The slate is still used today, even on the biggest movies. Why? Because it is almost 100% reliable, provides valuable information, and is very, very cheap.. Also (at the time of writing) a lot of material is being produced from multiple sources using Zoom or equivalent. There is no TC source so what is being done is people do a simple in-vision clap with their hands, and this gives a single sync marker – simple, easy, and about a hundred years old.

Big Tip: Never mix drop frame and non drop frame timecodes. If you need TC gear check out Ambient and Timecode Systems.

So, to sum up so far, to be a sound recordist you need:

- decent hearing
- enthusiasm for the job
- technical savvy

AND

Artistic Sense and Sensibility

How can I put that a little better (but still channel Jane Austen?)

Nah, forget it.

You are **not** just a technician making sure the levels are correct and there is no distortion.

Well, OK, in these days of increasing use of multiple tracks and radio microphones, maybe you *are* just a technician simply garnering the best sound for post production to sort out?

However, a sound recordist can (and I firmly believe should) contribute so much more in a positive way to the production as a whole. If you can't do that, where's the involvement and the fun? You have been turned into a techni-droid and your job will disappear eventually.

A recordist needs to think about the finished product and what we can do to give the production the sounds they need to make the right impact. An awful lot depends on the project you are involved in (and your motivation.)

Recording extra stuff is sometimes wasted effort – but you never really know.

I take a little digital recorder on holidays. Unfortunately to airport security it always looks like a taser.

Contribute artistically and you can't so easily be replaced by a machine

On a drama shoot especially you will often have spare time. This can be used getting atmospheres or wildtracks but before you all start shouting at me yes, I know a film unit is a noisy thing.

So consider giving your trainee a runabout kit and asking them to drive a mile down the beach to record some curlews. They'll love it.

And it won't do you any harm at all.

Quite often you will be in an enviable position to record sound that would otherwise be impossible or prohibitively expensive to acquire – migrating wildebeest, the atmosphere beneath a nuclear generator, a Lancaster bomber in flight, glacial calving, an author's last reading of his seminal work; an exploding volcano, a traffic-less Rome, and so on. Even if you don't need it at the time, most recordists will grab the opportunity to record an unusual sound or atmosphere. The best stuff, of course, is when someone forgets they are radio-mic'd!

You never know when it will be useful. At the very least you can whip out your smartphone as the pyroclastic flow thunders towards you. Remember not to scream, and *do* throw the device to safety at the last moment.

Who knows when you may be able to record something unique?

These days, there is little excuse for not pressing the REC button (unless you are out of power.) Once upon a time we had to worry about running out of tape or even recording over other stuff, but no more.

Far more importantly – especially in drama, documentary, music and NatHist, sometimes there is only one chance or one performance. It stands out. It is THE take and you should be able to recognize that.

Any director worth the name will also know it and they will want you to give them the nod; be it drama or opera, music or any sync related to wildlife, or some one-off event.

Of course, the circumstances are crucial – as are your decisions. Drama can be re-created. You may have heard technical faults (eg noises off, such as mobiles; crickets; or the 1812 Overture leaking from the next studio) but can that be mitigated?

Orchestras and operas and most performances can be stopped and go back and redo unless it's live.

Even wildlife repeats itself... lions, for example. But the question is - was it OK? Do you need another take? Is it even possible?

This is one of the most difficult decisions a recordist has to make. This is where you earn your money and reputation. No pressure then; (and what money and reputation I hear you ask?)

You will ***inevitably*** (*100% guaranteed* at some point) be asked by a director or producer:

"Was that OK for sound?" and it is not always an easy question to answer.

Sometimes the director is sending you non-verbal signals you should be able to read.

What if you reject it but post rescues it? (Bad)

What if you say it's fine and it isn't? (Very bad)

Maybe you play safe and ask for a retake? Is there time? Was that "THE" take? (possibly good or bad depending on whether the director needs an excuse but would like you to take the blame)

It's probably worth asking the director "Did you like that one?" to get some kind of clue.

I personally know of one recordist who had to admit, on being asked regarding a final "perfect" take (involving small children) if it was OK, that he had somehow failed to press the record button, both sound and vision, in the days of the Sony Betacam recorder. This happens more than you might imagine, but doesn't make it any better. Camera operators are not immune.

No, it wasn't me – My lips are sealed.

I am not sure if I have failed to record a take. Probably, is the answer. Or even certainly (if I was asleep at the time.)

As long as it wasn't "THE" take.

You never want to get a text from post saying: "We can't find slate XX take Z". Usually it is a technical problem - CF cards are ancient 1994 technology but are still used. Now we are on the point of streaming the rushes at the end of the day.

Or even continuously throughout the day.

That's gonna be fun.

As long as the problem is with the transfer, the rushes will be safe on your primary drive.

But none of that helps if you didn't hit REC

SPL

Sound Panic Level

Persuasion

How many people ever just stop and listen to the world? ... waves shifting pebbles on the beach; traffic clacking underneath a motorway overpass; a wheezing field of wind turbines; a storm brewing on the horizon; loud urgent next-door sex. No, forget I said that.

Listening – it's contemplative and relaxing and undervalued – birds, wind, traffic, planes, lawnmowers, leaves, laughter, tractors, (a washing machine at this particular moment of writing); are all common, but worth a listen.

And then there are the spine-tingly things like the Merlin engine, the Sunday morning bells of Florence, or a skylark high above the downs.

Lots of people enjoy a good landscape, especially in interesting light. They especially like pictures of themselves these days, usually grinning and doing something inane in their selfie with the Grand Canyon a mere backdrop, but it's rarer, I would suggest, to enjoy listening to a good soundscape, and if you can be bothered you will hear things others ignore – not because they are unable to hear them but because their brains just shut them out as "excess noise, not speech, therefore boring." If you don't enjoy listening to sound then you are quite a rare person because almost everyone enjoys music and that's just organised sound.

If even *that* makes you yawn then a recordist's life is not for you.

Most people don't actually concentrate on listening though - sounds a mite pretentious – doesn't mean it's untrue.

Something strangely wonderful happened early during the Covid-19 crisis of 2020. The human world went quiet. There were no planes. There were hardly any vehicles on the roads. Nature reclaimed the soundspace for a while and it was, frankly, though briefly, glorious. I am sure many, many recordists grabbed the opportunity to get their runabout kit out and take advantage, even if it was just in their own back gardens. Shame there wasn't any paid work around.

If you've ever done a production recce (see Appendix 3) you'll know exactly what I mean when other people simply don't hear the trains or the traffic, the planes or the air-conditioning.

Hardly surprising though as they are all concentrating on their own jobs. It's your job to notice and plan what to do about it.

Unfortunately for you, circumstances change and what was a lovely quiet location when you first visited now has an off-road vehicle driving school next door.

Ho-hum – out with the production chequebook.

Oh What... More Requirements?

Of course there are.

To be the best and most wonderful sound recordist you also need to be:

- a team leader

- able to operate as part of a bigger team

- capable of working on your own without any support

- an ambassador and negotiator for your production company or yourself

- organised for the ordinary

- prepared against the unexpected

Yes I know that all sounds a lot like motivational-bollox-speak but I am proud to say I made it up all by myself (which just goes to show how easy that crap is.)

Put another way (and less pretentiously) -

it means you need to be a nice person that gets on with people. You should be calm, organised, technically proficient, happy to work independently, and yet head up a team. The best of the best are fun to work with *whatever* the circumstances. I have to admit, sadly, I am not much fun in the rain. Also I cannot stand anyone playing with their smartphone on set, killjoy that I am, although that is very much a lost cause.

You will be responsible for your crew; from initial choice to their safety.

Sometimes you can work with who you want – other times you'll have to work with whoever is hired by production. You'll usually have some input. Many recordists work closely with their own crew with whom they have developed a years-long relationship, but people do get other offers, or get Covid-19, or decide to move on, so you have to be flexible.

There will also be a lot of planning involved regarding:

kit, transportation, suitable clothing, communication, accommodation, accreditation, immunisation... the list goes on.

You are Head of Department and therefore responsible for liaison with Production which means you must attend recces, production meetings, safety meetings, technical meetings and complete all the certificated courses that the production requires before the shoot begins.

This is a lot of pre-production work, and masses of emails and calls. I mean – hundreds.

Has the boom-op completed the mandatory cyber-security training? Is anyone in your department allergic to meerkats? Have you passed our "Driving Safely between Locations" skillset? Do you have a tickly cough? Please prove who you are. Please amend your risk assessment with regard to the latest pandemic conditions.

It's amazing anything gets done.

Especially as (and it WILL happen) there are last-minute changes that affect your plans, health and safety, and your hopes for an easy life.

As a valued team member, however, despite your heavy responsibility and exalted position, you are still only a small cog in a big machine. The larger team is the production as a whole. At least, let's hope it is a team and it's on your side. If all is running smoothly then you only need worry about your actual job and not all the associated poo, and all will be well with the world. Sometimes though communications do break down and the results can be expensive.

Try and make sure it's not down to you.

If in doubt, before the shoot, talk to someone, even if the production company is on the other side of the world

(though it's only polite not to wake them at 3 a.m. local time - they will be mightily pissed)

If you have technical questions I recommend talking to somebody technical like, well, yourself. They can tell you about frame rates and software versions and post-production requirements and nobody from production will be of any use (why should they? – it's not their job) but production can at least give you contact details if you know who to ask for.

If you have organisational issues then talk to the production office.

It's common sense to try and avoid problems before they occur and that way we all keep smiling (though it may be wise to steer clear of those with slightly glassy eyes who are constantly grinning despite the godawful circumstances. I have seen a few like that. You just hope they get better before they implode.

Or worse.)

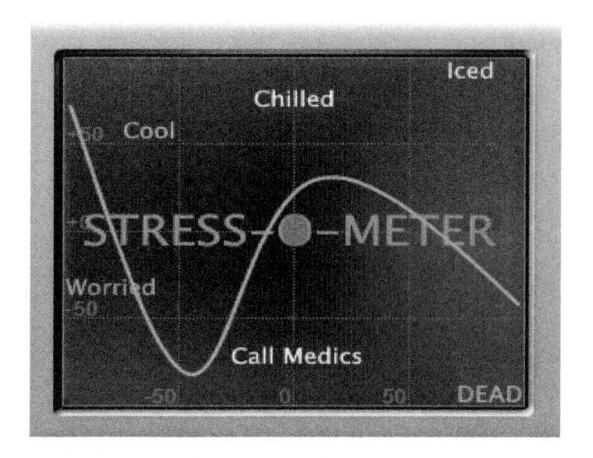

Working Independently

Recordists sometimes need to be able to operate alone and without immediate direction. Motivation, discipline, common sense, preparation and a set of defined goals are goodly things. Once again I slightly cringe at the language I have used but it's all true. It's nice to be in a team, but it's even nicer to be set free, on occasion, if you enjoy that feeling.

Hypothetical example -

> *A prestigious wildlife series is nearing completion of post production. The wildlife photography has taken 30 months. Most of the esoteric calls and sounds are available except for a particular bird in the Brazilian rainforest because the camera mic failed in the humidity. Scraping the bottom of the budget-barrel, the production advertises for a recordist to go out and get that call. They will pay receipted travel, accommodation, expenses and equipment hire, but only a nominal fee in case it's a washout, and you have a week to do it.*

How would you feel about accepting that job? Would you negotiate a bonus for success? Would you suggest you take a small HD camera just in case? Have you got all the correct jabs for Brazil? What language do they speak there anyway? Do you know *anything at all* about birds?

Your answers will give you some clue as to which direction your career should take and how many untruths you should or should not tell, because it *could* all come back to haunt you.

Usually the production will have someone very specific in mind that they want to do the job, but that person is not always available (especially for a mere week and at cheap rates) and that's when they get desperate. They really want to use somebody they know and trust.

Using freelancers is a double-edged sword.

This could be the opportunity of a lifetime, or a one-way street to Shitesville.

Suppose it was a job to record an exploding volcano – how would you feel about that?

The cathedral/violin recital is looking like a pretty good option now isn't it?

Welcome to freelancing.

The Exagge-Rate Plug-in

Experience

"Good judgement comes from experience, and experience comes from bad judgement." Rita Mae Brown

Experience means you have encountered something before and are now more ready to cope with it. You learn from both good and bad experiences, and it's the same for most jobs. Learning from bad experiences is hard – and yet it teaches you the most. If you can learn from others' bad experiences – all the better (kinda no-lose really)

Out of fuel and out of communication in the Masai Mara in Kenya I got the producer (the tallest person) to stand on the roof of the Landrover holding the vehicle jack above his head on which was attached the magmount ariel for the walkietalkie. My experience with Outside Broadcast communications helped here. We got through to base, and were rescued, though we felt exceptionally stupid. It was the bloody producer's fault though because he was driving. No names, but what was that bollocks about a dodgy fuel gauge John? It wasn't really a serious problem because the Mara is stuffed with wandering vans full of tourists and spare jerrycans.

Why didn't we have any emergency fuel? I don't bloody know I'm just the recordist ff's sake!

But really experience means you are more likely to cope with unexpected circumstances, and there is no substitute (unless your experienced team pulls you out of the shit, unlikely though that is, because a recordist is never wrong)

As a boom operator I used to shoulder far more responsibility than I should and I didn't always have the seniority to deal with it and got into trouble a few times. Ah, well, that's inexperience for you.

I tell my assistants and boom operators to always do what I advise them if they are having a problem. That way, they cannot go far wrong. If they do what I say and there is a further problem then it is my fault. But I am old enough and thick-skinned enough to take the flak.

It is blindingly obvious but quite amazing how many people do not appreciate this. On the other hand -

I kinda lost my temper with B-camera on a drama and stomped on to the set and harangued the operator for doing something stupid and unplanned. I don't recommend this approach.

The Director took me aside afterwards and made it clear she didn't want that kind of shouty-thing on her set. She was quite within her rights, of course.

The problem is that B-camera operators desperately want to be A-camera operators and, in their enthusiasm, will try almost anything to prove their worth: not thinking about anything else like, for example, zoomed-in dialogue that is not being covered by sound, or the fact the artist isn't even lit.

Their shots are worthless and the editor wonders what the hell is going on.

But I suggest, having had this experience, it is better having a private word afterwards and explaining the problem. You sometimes have to do this a number of times with particularly dense or particularly fervent B-cam people.

Or you could do what one well-known recordist with a fearsome reputation did and march onto set, go up to the B-camera, and, with an appropriate expletive, switch it off.

Experience lets you be better at your job. You might not be able to jump over tall buildings but you can hum a little tune to yourself whilst the director is having a 5 minute rant, and then **calmly** sort out the problem.

What problem?

Exactly.

Experience helps you to rapidly diagnose faults and either get them fixed, or to think of another way round. It teaches you what you can and can't get away with, when to worry, and when you can chill.

Eventually, the apprentice will become the Master.

EVIL CHORTLE

"I thought I made a mistake once, but I was wrong."
T.H. Production Sound Mixer

5

MAXIMUM GEAR

SOUND KIT

THE OBVIOUS STUFF YOU NEED

KIT - The sound trolley/cart

Recordists are famed for their esoteric trolleys (known as carts in America and elsewhere). They represent years of accumulated wisdom made real in an organic agglomeration of pieces of kit, wires, unfathomable (and usually entirely redundant) equipment, switches, secret drawers and flashing LED's, and no one sound trolley will *ever* be the same as another. It's a kind of universal rule, like: *if you do not love your cart you are not a proper sound recordist.*

Sorry, but that really is a universal truth.

Of course, you can take things too far; it's all a matter of taste and preference.

Recordists who enthuse about their latest fixtures and amendments risk being boring farts because they cannot see that it is, in fact, of no interest to *anyone at all* (except other recordists who will cheerfully nick their ideas at the drop of a tweaker.)

It's all a bit different to camera department who always seem to have the very same "must-have" bits of gear piled up on the groaning and consistently overloaded magliner.

But if you want to be a recordist you will be keen to know the secrets of the perfect trolley/cart...

First off; spoiler alert – there is no perfect. However, there are some important points (some obvious, some not, some deeply secret.)

You can build your own cart from scratch, or choose from an ever-expanding, wondrous variety:

>http://www.soundcart.tv/

>https://www.raycom.co.uk/product/soundcart-production-sound-cart/

>https://www.locationsound.com/carts-24

The cart needs to contain, in an accessible way, all of this:

- Recorder/mixer

- As flexible a monitoring distribution system as you can devise

- Radio mic rack

- RF extensions (to both Rx and monitoring Tx)

- Antenna holder(s)

- Boom holder(s)

- Light source(s)

- Storage cubby holes (as many as possible, but even that will be inadequate.)

- Cup holder – I had one from a SAAB (donated to me by a lovely script supervisor) but even *that* broke under filming duress so I suggest you source the toughest one you can find and screw it on so that it is convenient for a quick sip and yet will not spill onto anything/anyone important in the event of a mishap.

- Monitors – quite important for drama especially. Unfortunately they take up a lot of space and power and you need to keep up with the distribution protocols which change ad nauseam due to the progression of video standards. Newer video streaming hardware is being introduced all the time, allowing anyone on the crew with a smartphone to access the image from the camera. This really is the future of on-set video distribution.

- Comms – you need separate comms with production and your sound team and sometimes a third party (or more.)

- Script and notes – old-fashioned paper has a number of advantages, but so do notebooks – just remember whatever you use must be weatherproof and never run out of power.

- Powering – please think of your crew – you, of course, love big batteries, but the crew hate them. One day technology will solve this problem. Maybe. Trouble is, the more power is available, the more you will use it up. If you are on a low-loader or up a mountain or on a boat or in a balloon and you run out of battery power you are going to wish you were dead or, at least, in a very deep hole, (unless you just happen to have that camera-battery to four-pin XLR with you, and a pocketful of dry cells. More weight, of course, but better than hoping to be struck down and put out of your misery by sudden myocardial infarction.)

Sometimes (well, quite a lot of times actually) you'll need to access the inaccessible (location-wise), so breaking down and then reassembling your cart should be as easy as possible.

(Mine un-docks from its wheels so it can be strapped into a seat – car, train, plane, airship, rocket. Ha! Bet no one else thought of that!)

You'll see now why there is no perfect design. Every recordist will insist their cart is best. Of course they do; it's best for them and, with luck, no one else will be able to figure it out, especially that annoying up-and-coming assistant who asks rather too many pertinent questions, or worse, questions to which you have forgotten the answer.

Recordists enjoy designing their carts and then, in any hiatus between jobs, reinventing them with the benefit of experience. I should be doing that now, but I am writing this. My trolley has a hand-carved wooden drawer on heavy duty ball bearing triple extension runners to house the flat bed mixer. It has the same satisfyingly solid feel as a chunky quality button in a Lexus or BMW. I like to open and close it and feel those silky bearings rolling silently in and out... I should stop now shouldn't I?

KIT - Microphones

Intrinsic to the sound kit, obviously, is the microphone. It picks up the vibrations in the air and translates it into an electronic signal that can be amplified, and then you can do what the hell you like with it. They vary from the stupidly cheap to the mind-numbingly expensive, but they all do the same thing, though in a variety of ways and qualities.

The actual worthwhile makes are very much a matter of personal preference, peer pressure, fashion and advertising but you generally get what you pay for in terms of ruggedness, reliability and output quality. I would urge everyone **not** to go for cheap unless it is disposable and likely to get trashed (in water, or a stunt, for example, and production have agreed to cover the cost.)

Example Drama microphone loadout:

- personal (hidden, various colours, windshields, etc.)
- close vocal (in vision)
- spot
- gun
- omni
- stereo
- 5.1 or other ambient

In almost all circumstances, the closer you can get the microphone to the desired sound the better. However, this is not always the case with high transients, very high spl, or mics prone to bass tip-up.

Magnificent Microphones

How about the industry standard Shure SM58, in production since 1966 and still going? Did they get that right or what? Well, actually, they now use new adhesives and materials and manufacturing processes but the specs are very similar.

The SM58 is very much an in-vision or stage mic (SM stands for studio microphone) and is not much used by recordists but by recording engineers (especially vocals) so let's look at another classic –

The Sennheiser MKH 416 short gun

This (and its antecedents) was a really clever microphone design – it was low-noise, directional, and could be used in humid environments (like the open air!)

The history of the MKH series is fascinating. It begins in the mid 1970's, and although the 416 (and its big – more directional - brother the 816) are no longer manufactured, they were industry standard for 50 years.

The MKH name continues and I use the 30, 40, 50, and 60 series of Sennheisers. They have proved to be reliable, reasonably tough, weatherproof and RF resistant, and I love the natural sound quality. What they will not tolerate is a fall from an ill-placed boom onto a hard surface.

Nudge – buy a bigger foamy – the Sennheiser-supplied foam windshields are totally inadequate. Maybe your microphone comes properly equipped? Test it, as if you were a boom operator matching a fast tracking shot. Absolutely no one wants to hear that they have to go again because of wind noise in an interior set.

Other excellent manufacturers to consider - Rode Schoepps AKG Neumann Sony

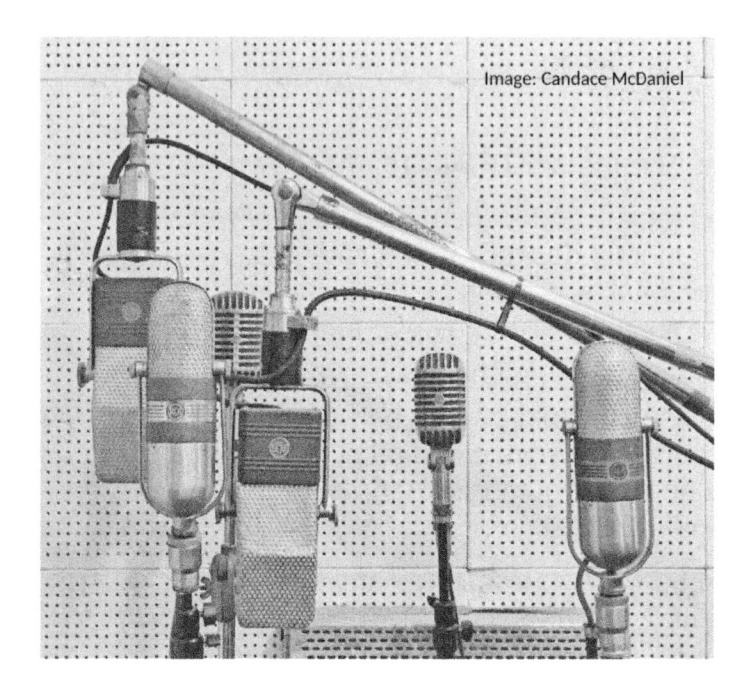

Image: Candace McDaniel

KIT - Mixers

Once upon a time there used to be portable mixers (such as the SQN) and separate recorders (tape, DAT) and a whole load of wires between them but, hurrah, now we have solid state, combined devices.

It's hard to imagine what it was like when recordists all used the Nagra quarter-inch tape machine in the field. It was bulky, heavy and each reel of tape lasted a mere fifteen minutes.

However, it was beautifully made, rugged and reliable. I know of at least one recordist who dropped one in the sea and managed to recover it, thoroughly hose it down, dry it out, and power it up again (I don't think I would have bought it off him though.) The Swiss firm is still making expensive, jewel-like and exquisite machines. The less-well-known alternative was the Stellavox which still has its advocates.

Now each year brings us a shiny new digital recorder/mixer from Zaxcom, Fostex, Zoom or Tascam and at shockingly reasonable prices.

They all record onto some type of SSD and onto removable cards. You can upload to the cloud. You have routing matrices, EQ, compressors, TC generators, faders and hundreds of hours of recording time all in a convenient box the size of a large to medium tuna sandwich.

Which should make things terribly easy but, like all kit, especially the mixer/recorder, familiarity and experience is everything.

KIT - Radio mics

Most recordists have a love/hate relationship with these devices. Once again, the more you invest , the better they will be in terms of reliability and ruggedness. There are some situations where you simply cannot do without them – the massive wide angle; the ship-to-ship shot; the presenter dangling over a cliff describing the sex life of the Guillemot; the list is endless.

But they are also an endless source of worry, frustration and heartache.

They can suffer from interference, dropout, clothing noise, wind blasting and, worst of all, sudden battery failure even though you fitted a new one half an hour ago – yes, you got yourself a bad batch but the voltage appeared fine when checked.

They are all limited in power output due to government regulation so they all have a limited range, but that range differs according to the weather, how much the wearer is sweating, the frequencies you are allowed to use, the density of other RF traffic and just pure chance. It's a black art, and I have only met one or two people who fully understand it but they were, umm, bearded technical types unused to human company who enjoyed warm, dark real ale, and sexy talk of half-wave dipoles, antenna ground planes, and heterodyning.

However, new technology is coming to the rescue. Reliability and range are increasing, and the risks of dropout and interference are being mitigated, whilst still complying with the maximum output limit. Some of the new kit is amazing – though hardly cheap. If you are starting out, it makes sense to hire radio mic kit appropriate to the job. Eventually, though, you are going to want to own and maintain your own setup.

Get the best and most modern you can possibly afford, and you'll be glad you did.

Examples -

> Audio, Lectrosonics, Micron, Sennheiser, Sony, Wisycom, Zaxcom

KIT- Sound Distribution

This means, in essence, providing a load of headphones plugged into radio receivers for people who want to listen to the dialogue, or the sounds you yourself are hearing. It might be just the director. It can be an awful lot more complicated than that, but that will usually be in a studio situation, and the sound supervisor and his mixing desk will easily cope.

On a location drama, what you will most likely be expected to provide is a monitor mix for the director and the script supervisor. Alas, there will also be a whole legion of others wanting to listen-in, including several AD's, the producer, friends of the producer, advisors, stunt supervisors, child actors, their chaperones and, probably, Uncle Tom Cobbly's dog-minder. Oh, and actors quite often want to review their performance, and ruggedised pvr's are now popular so they will need a feed as well.

The airwaves are getting well crowded, so you will get complaints of interference. Also, your kit will be treated with absolutely no respect and pulled apart, split, broken, forgotten and dropped

in the toilet more often than you can possibly imagine. It hurts. Your poor crew will waste a lot of time on location trying to keep up with who has walked off with what and, all too often at the end of the day, they will show you the mangled remains with a shrug and: "the caterers found this on the bus."
Hint – keep your receivers bagged and insist they remain in their pouches (this can prove very difficult with some people.) It is impossible to have too many spare headsets, but then you start to make a rod for your own back.

I once thought I would buy and keep a *secret* emergency spare prior to a shoot.
It was in use by day 2.

KIT- Bags and Cases

Aluminium travel cases are customisable and well worth the considerable outlay as they will last for years and protect your precious gear. If you dream of travelling the world you need to spend money on the right cases but you will be doing yourself a favour.

You may be tempted to try using "tough" suitcases or such but – one, it doesn't work, two, it looks unprofessional. Sometimes all you need to do is look at the dents.

The last thing you want to do upon starting a shoot is be getting on the phone to arrange replacement kit damaged in transit.

It depends how you like to work.

Some recordists want to take everything to every location. All very well unless the location is a cove 200 meters down a narrow cliff path. Or up a mountain. Or in a balloon. But then, you never want to appear like an idiot who never even considered needing half a mile of cable on a beach or a desert. Or wherever.

I can understand if you want to take everything, but I also sympathise with the poor crew who secretly hate your big, sturdy cases.

KIT - Bits No One Thinks About

I suppose I could try to give you a comprehensive list but that would be boring, and I'm bound to miss a lot - "Oi, what about an entrenching tool!" you would suggest, and you'd have a point.

You simply cannot have enough convertors of – well – everything to everything else, though I would stop at mains to XLR because someone will likely be electrocuted, and it'll be your fault.

Then there are:

straps, bungees, plastic ties, heavyweight plastic bags, ground sheets, carpet, umbrellas, wooden wedges, cleaning materials and sprays, brushes, full electrical toolkit, microphone mount spares, specialist tools, torches, tyre pump, padlocks, keys, jump leads, swiss army knife, spare sharpie pencils and notebook, gaffer tape, toupeé tape, self-amalgamating tape, personal weatherproof kit, blue-roll, antiseptic, F50 suncream, and comfy fishing chair. Possibly, an entrenching tool for the really worst shoots; you know, the ones without latrines.

If you are still interested you can find my list of electrical bits and pieces of incalculable worth in Appendix 1

But experience is the only thing that will teach you what you forgot. Loperamide, for instance.

KIT - Pride and Prejudice

It's a sad thing to say, but a lot of important people will be impressed by the size of your equipment.

A few production-types know a bit about camera gear but I have yet to meet anyone "higher up" who understands the slightest thing about sound thingies (with the single exception of a director who used to be a sound recordist – but he was a joy to work with so no complaints there. Anyway, come to think of it, directors don't count as "higher-ups" as they probably work too hard.)

Nudge – some recordists might be tempted to add extra equipment, especially stuff with flashing LED's, that doesn't actually do anything. Some recordists might use a mixer that is far bigger than they need, but rows and rows of faders are mighty impressive These recordists might point out, if cornered, that they need redundant systems in case of failure. They would be right, of course.

Everything a sound recordist brings onto the set is totally necessary, and who is going to argue? These days you can buy some wonderful surround sound metering with bright OLED displays to show everyone how clever you are for a mere few hundred quid.

Even better are big modern radio mic racks, but they are several thousand precious monies I'm afraid, but generally worth it. Make a good show, and others will always be impressed and say to each other "we must employ him (or her) again because we are obviously getting a lot for our money."

I was booming (many years ago) for Dicky Bird, a lovely, self-effacing recordist, and a joy to work with. He had nothing to prove but even so the producer appeared on location one day to do the rounds and actually complimented him on his impressive trolleyful of equipment. I couldn't believe it – he was only using one bloody channel – the one connected to me and my boom. He just winked, and later explained to me some of the realities I had so far failed to comprehend.

Throughout my career I have always tried to be fair with production managers and the money people – too fair probably. Sound kit hire costs next to nothing compared to camera gear and grips' equipment and the design budget. This is because it does not depreciate as quickly and doesn't get such a bashing.

In fact, some older ribbon microphones are in such demand that they are far more valuable today than when they were first manufactured in the 1920's. You can't say that about video gear!

KIT – The Secret Stuff

What? You honestly thought I would reveal that here?

Oh OK then.

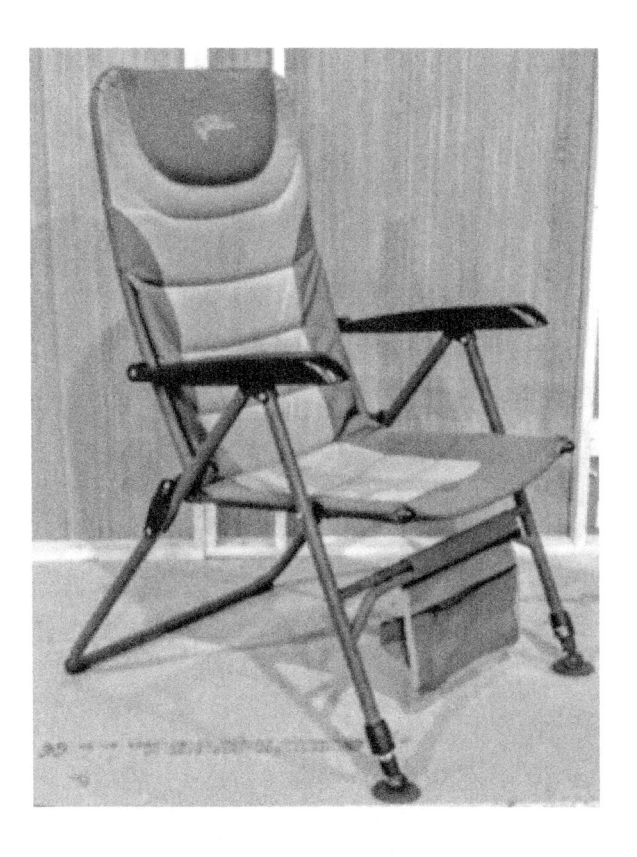

The secret involves a good chair.

It must be tough, light, waterproof, foldable and yet comfortable and adaptable.

It must be the perfect height for you and your equipment.

Sometimes sound assistants forget to bring the chair onto set or location, but they only do that once or twice at most, or they mysteriously disappear.

My chair has three positions - work, relaxed, and lounger; the last position is for sunny exterior lunchtimes only. Relaxed is for rehersals. A good chair engenders chair-envy but you can put this to use and invite attractive costume and make-up assistants to give it a try. You could also invite unattractive sparks or grips to give it a try - up to you.

The chair is probably second only to the cart in importance regarding sound kit. You can add things to it - here I have attached a waterproof bag for batteries, pencils and wrapped snacks. Whatever it is, (battery or snack) it should be easily identifiable in the dark with fingers only. It is unhelpful to confuse an AA battery with a Twix^{tm}, for example.

Inserting a Twix^{tm} into a piece of equipment is actually worse than munching an AA, but only just. Maybe it is worth keeping a torch in there?

Or consider a change to Liquorice Allsorts^{tm}

Really Big Kit

And the winner of biggest boom shadow is ...

Really Big Cat

Anyone spotted the chocolatey theme yet?

6

FINE BALANCING

HOTGO or "HOw To Get On with..."

The Sound Team

The Sound Recordist is Head of Department on a shoot. That means you take all the credit and distribute all the blame to anyone you can think of. No it doesn't.

If you get a reputation like that you are sunk - I certainly hope. Your aim should be that you are so good to work with and your reputation is such that the best of the best will fight to work with you, or hire you so - happiness and reflected glory all round.

Besides, the whole idea is to enjoy your work, and if your team are not enjoying their work, you have failed in some sense. It might be that it is just a rather sombre, unhappy shoot. It happens, but you need to mitigate that if you can.

I cannot over-emphasise how important a competent, integrated and happy team is to you. It relieves stress, and leads to a better overall delivery stream. Which is bollox-speak for "it works for everyone if everyone is working together"

Or something.

Motivated smiley people are generally more pleasant than dour and lazy dickheads.

It's pretty obvious, though I have to say it's actually rare to come across dour and lazy dickheads because they generally don't last five minutes in the business (I can only really speak for sound – some departments have their own special problems.)

If you don't agree with running a happy team then happy heart-attack.

I want a crew who are experienced and on-the-ball, and know when I need that cup of tea. It is a joy and a pleasure to work with such people. One day I joked about where the sound doughnuts were. Next morning they were there! Ace crew.

(note to our American friends) – we don't usually have craft services this side of the Pond as productions are too mean. We might have hot water and a packet of cookies between a hundred people. There is catering, of course, but no special treats. I don't know how we cope to be honest. Mostly, I suspect, by talking about how delicious doughnuts would be, if they were there.

We absolutely love it when you come over here and bring your productions, your realistic money, and craft services though we do get a lot fatter.

It's happening a lot now, and at the time of writing, the UK is becoming a cool place to shoot.

HOTGO

The Director

"I never said all actors are cattle; what I said was all actors should be treated like cattle." Alfred Hitchcock

"Where's the emotion? I wanna hear it!"

You obviously need to develop a good relationship with your director. I have worked with hundreds, ranging from the intense, the wonderful, the dreary, and the downright pop-eyed barking.

They have a very demanding job and they can't stop from first call to wrap (and before and beyond) and everyone wants a bit of their time.

You are definitely not uppermost in their thoughts.

Unless you have worked with them before it'll take a couple of days to learn how to handle them.

To avoid any possible embarrassment, here is a guide, and a warning, regarding the archetypes you might meet:

The Newbie

> Needs TLC. You all need to pull together as a crew or you'll never wrap. He or she knows the theory but if this is the first time on set they will either (a) brazen it out or (b) ask for some input. My advice is to be kind and take a little bit of time to explain why you are doing what you are doing or why this particular thing is a problem, but don't be pushy. You could be doing future generations of recordists a favour. After all, who knows where this novice will end up? Head of Drama probably. Hopefully all the other crafts will be doing the same as you, especially the DOP and Operator. It need not be painful, unless the novice is insistent they do it *their* way. Try not to roll your eyes if this happens. Be polite, professional and helpful and never condescending.

The Old Hand

Very often the best type – they know what they're doing and likewise expect you to know your job. As long as you don't f*ck up they'll leave you to it. They can range from absolutely lovely and charming to curmudgeonly old scrotes but you'll likely wrap on time every day. The nice ones will often give you fair warning of any problems they foresee. Others might just take slightly evil and silent pleasure in watching you cope (or not cope.) If you do f*ck up they might not even say anything – you'll just feel horrible, like you've let them down; which you have. Enough said.

The Dictator

Usually it's the Operator who suffers the most but sometimes the Dictator turns their attention to another department to keep them frightened and controlled. Every now and again it's Sound's turn. They may have an idea of how a scene should be covered that is different to yours. They may hate radio mics. Or love them. They may insist on overlapping dialogue, or running playback music over dialogue, or have characters whisper in a working train station. They are a nightmare. Most are bullies, and like bullies they have favourites. If you are a particularly ingratiating type you may well charm the Dictator and they will leave you alone. You have to judge when to stand up to them and when to acquiesce. It is unusual for a sound recordist to walk off set after a disagreement with the director but it has been known. The Dictator quite often reduces people to tears, and enjoys it. The thing to remember is – it's only entertainment; none of this is real, or actually matters. The Dictator is really a child throwing things out of the pram because they didn't get quite what they wanted. Sometimes Production will be supportive of any complaints; sometimes not – depends if the Dictator is flavour-of-the-month. You'll either put up with it or you won't – but if you do throw off your headphones and walk you will probably be respected for it in the longer term. You'll certainly make a name for yourself.

The Artist(e)

A director who wants to achieve something no other director has ever conceived – they are obsessed by camera angles, or the nuances of actors' moves or lines or the colour of the settee. Generally speaking they'll ignore you completely. Go with it, but be prepared to get back to the hotel each night rather later than you hoped. Nobody else will understand the difference between take six and take twenty six either, so you are not alone. You will be professional enough to remain stoic, and practise daily a Zen-like acceptance which will serve you extremely well for the rest of your career.

The Speed Merchant

They range from the fizzingly fun to the sizzling shite but they all like to work fast and if you can't keep up (or better, get ahead) you'll be off the shoot. Quite often for performance reasons they'll only do one take. If you're lucky you'll get a rehearsal. These guys and gals are not good for your blood pressure but you might get to wrap early and you *never* get bored. They are, in my opinion, the most exciting, challenging, and fun of the archetypes, but you need to be on top of your game. Watch out for distracted boom ops or assistants who haven't quite realised what is going on and give them a pep talk (aka a boot up the arse.)

I really enjoy working with this kind of director as long as they have a sense of humour (especially you, Jordan.)

To sum up, a good director will:

- trust you to do your job to the best of your ability

- listen to legitimate problems and help you solve them

- give you some feedback if there are any concerns (technical, artistic or regarding personnel)

- inspire you and make you want to work with them now, and in the future

The above doesn't just go for you, the recordist (what are you, special or something?) I have a lot of admiration for (nearly) all directors – they must put up with pressure from above and below. They must deliver a project that is of quality, on time and within budget. Most of them will be thankful for anything you can do to help with that. That may involve compromising, or making that extra bit of effort which you will want to do if they have let you enjoy your job and do your utmost. The best will inspire every craft in the same way, and you'll be involved in a happy and rewarding shoot.

Honestly, if you like and respect them, why wouldn't you want that?

Of course there are a whole load of pleasant jobbing directors out there who are content to concentrate on their job and get the thing shot and edited, and then move on to the next project. They don't *really* care who they work with as long as everybody is professional, efficient, and (an added bonus) enjoyable to work with.

Not an unreasonable expectation.

HOTGO

The First Assistant Director

The 1st AD has nothing whatsoever to do with directing. Their job is to get the day's schedule done (impossible though it might seem.) They are basically organisers, and they have a lot to organise.

They are always the loudest voice on set. They can be your friend, or vilest enemy.

Not all 1st ADs care about sound in the slightest and you may have to fight your corner. Be aware that *their* primary concern is moving things on so that a wrap can be achieved on time and budget and they don't like delays (because it makes them look bad.)

The 1st will either ask at the end of a scene if sound is happy (you have lucked out here – this is going to make the shoot so much easier) or they won't.

Nudge – Be nice. Be understanding, but tell the 1st what you will need before the end of the scene. They *might* even remember. In certain circumstances (when ADR is not an option and a background noise springs up, for example) you may have to insist that shooting is suspended so that you can get a buzz track. This should be done through the 1st and with the agreement of the director.

Sometimes the 1st may overrule your request for a W/T, atmos or whatever. Don't get upset – there is usually a good reason – like the location venue is going to double its fee or something. If there is not a good reason, then just put their name down in the "List" with appropriate comments (e.g. unsympathetic dickbrain; avoid.)

Remember the job of the First AD is to get the Director's wishes done in the time available, so if you can somehow make *your* wishes the *Director's* wishes... you see where I am going?

Nudge – some sound recordists like to be close to the "village" and some don't – preferring to be secreted away where no-one will disturb them. I find it is better to be with the Director and DOP; you can listen-in to their plans, and be one step ahead. It doesn't work all the time by any means. but you can always push the idea of wildtracks or shooting in cloud, or calling lunch when the local schools empty into the playground.

Video Village – survival of the fittest. It's a jungle in there.

This is a really good time to get those mints out – the ones you keep in your back pocket for dire emergencies

HOTGO

The Director of Photography

Q: "What's the difference between God and a DOP?

A: "God doesn't think he's a DOP"

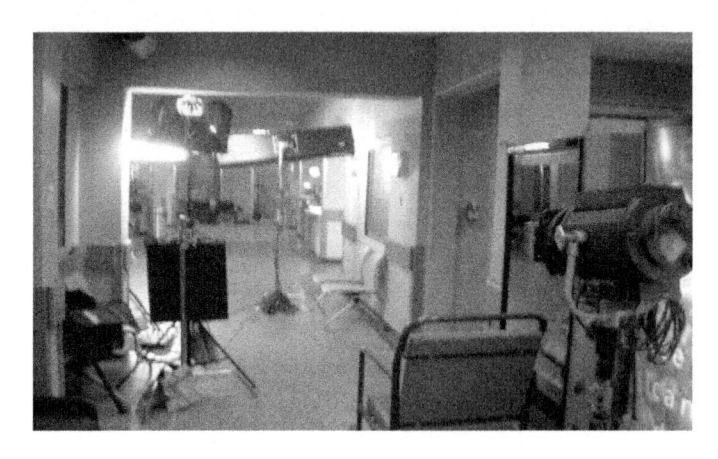

You can separate DOPs into almost as many groups as Directors. The Old School Prima Donnas are the worst. They will be cheery backslappers happy to regale a crew with anecdotes from yesteryear in the hotel bar but they are totally inflexible regarding their lighting. Work around it – or suffer. Luckily they are dying out. Yeah, I know that sounds harsh.

However, the vast majority of DOPs are perfectly reasonable and will work with you to solve shadow or other camera-related problems, but, in the end, their top trump beats your top trump so it's advisable not to forget that.

It is with a heavy heart that I admit that lighting is rather more important than sound. Both can be fixed in post but, unless you are re-engaging an A-lister for ADR I reckon sound is cheaper to fix.

The DOP's lieutenant is the Gaffer – if you can strike up a relationship with him (more rarely her) it will ease things like getting a noisy head or ballast changed on set, or even moving the generator(s) further away. Good luck with that one, by the way. Luckily, technology is once again coming to the rescue regarding noisy lights.

HOTGO

The Camera Crew

You'd be a fool not to try and maintain a good relationship with this lot. It's not always possible – sometimes they may be a bit aloof, but mostly they are so shit-scared of being shouted at by someone they haven't got time for pleasantries. On smaller crews it is easier, and indeed almost essential to be on good terms with the Operator – a nice one will adjust to keep the boom out of shot but don't push it too far. It is very, very easy for them to make your team look amateurish with the merest tilt.

Your assistants need to interact positively with the camera crew especially during rigging, asking permission to rig TC receivers or onboard mics, for example (essential with Steadicam.)

There is rarely any problem with the Camera Dept. as we are all on the same side in the end. However, mavericks can upset things, and, as always, personality is everything, and this relates directly to your sound assistants.

Camera and sound crews quite often socialise after work and this is no bad thing, An awesome, fun camera crew both on and off set is a joy (alas, too many names to mention.)

HOTGO

Costume Department

This relates mostly obviously to drama, but not exclusively. Do not make the mistake of thinking that, say a Medieval Science Fantasy scene will be more difficult than a modern Emergency Rescue scenario because it all depends on the materials used in the costumes. Costume dept. can help by sewing in special pockets for transmitters or little secret places for microphones. Don't forget their remit is not to make you happy – it's to make the production happy. The on-screen appearance is, frankly, more important than your sound (to a costume designer) but they are usually aware, and they will help if they can (and you are nice to them and buy their ice cream) but sometimes a Goretex jacket is a Goretex jacket and nothing can be done about that plasticky rustling noise.

They might also be able to help your team with

clippy-cloppy shoes, though high heels remain, and always will be, a sound nightmare. Sometimes new and keen costume assistants can be a pain, but they are only trying to do their job – i.e. making sure your bloody radio mics can't be seen in shot.

Make sure your team are aware of this. It's all about compromise. Nobody wants to see the bulge of a radio transmitter or a sneaky lead on an artist when a drama is broadcast, do they?

By then it's too late.

Even on a documentary it's nice to have everything neat and discreet. Once again, there is no substitute for experience and a friendly personality because sometimes you are going to have to get quite intimate with people.

Actors vary from the pro - "Oh just stick it where you need to" to the vulnerable and unsure where rigging microphones is concerned.

It's a significantly underrated skill – you need to know where a mic needs to be. It will often need to be invisible, yet accessible for adjustment. Artists and costume department need to be happy with you doing what you have to do and attitude and people-skills are all-important. It is especially important when dealing with children.

The Covid pandemic presented particularly tricky situations from which we have all taken valuable lessons.

HOTGO

Conversations on set (decoded)

Dir: "How was that for sound? I thought I heard a car/plane/telephone/cough"

Meaning – I need an excuse to go again - please give me one.

Dir: "How was that for sound? I know there was a car/plane/telephone but..."

Meaning – I really liked that take. Please tell me we can use it.

Dir: "I'm going in for coverage."

Meaning – don't worry about the super wide-angle.

Dir: "I'm not going in for coverage"

Meaning -"Get the radio mics out"

Dir: "Was that overlap OK?"

Meaning – I liked it – can it be sorted in the edit?

Dir "Was that overlap OK?"

Meaning - should we do that again? – the editor is giving me a hard time.

DOP: "You're not going to like this."

Meaning – I have totally and knowingly screwed you with this lighting setup. Deal with it.

DOP: "Oh I suppose I can turn that one off if you insist."

Meaning - I totally forgot that lamp was left on from the last scene and it looks better off anyway.

DOP: "I think there may have been a boom shadow on that one."

Meaning – There definitely was a boom shadow and it was so bad even the viewers will notice.

Art Director: "I think there may have been a boom shadow on that one."

Meaning – There may have been a fleeting shadow that nobody else will ever notice but I thought I'd point it out anyway to show I'm awake.

1st AD: "Anything for sound?"

Meaning – do you need any wildtracks, buzztracks or re-recording of any kind before we wrap this location? Oh, and you owe me a beer.

1st AD: "Anything for sound quickly?"

Meaning – I am pressed for time. Do you *really* need any wildtracks, or such shit?

1st AD: "Moving on!"

Meaning – I have totally and purposefully forgotten any requests from sound department, or indeed, anybody else.

Camera Operator: "The boom dipped in at one point."

Meaning – The boom operator made a mistake and dipped into shot.

Camera Operator: "I think the boom dipped in at one point."

Meaning – I inadvertently tilted up and caught the boom in shot, but I'm not admitting it.

DOP: "We can probably squeeze/paint that out."

Meaning – I'm bored of this scene anyway. You owe me a beer.

Sound Recordist: "Sounded like we hit something."

Meaning – if you break my microphone you incompetent boom-tit I'll kill you.

Sound Recordist: "Couldn't hear the words over the Wildebeest."

Meaning – this is the worst tracking shot ever. Somebody shoot me. Or preferably the camera crew.

Focus Puller: "That may have been a bit buzzy"

Meaning - I rather screwed up the focus and we need to go again. Sorry everyone.

Costume Assistant: "There's a small continuity error."

Meaning – We just realised the artist is in the wrong costume for this scene.

Stunt Co-ordinator - "After the explosion, the vehicle should end up around here."

Meaning – Be very afraid, and give yourselves an extra 50m safety margin.

Special Effects Supervisor - "You may want to wear these ear defenders"

Meaning – Oh dear God this is going to be fucking loud. Where the hell do I set the fader?

Location Manager - "We can't shut the construction site down today."

Meaning - We haven't got enough money to shut the construction site down even for one day.

Location Manager - "This street can get quite busy during rush hour."

Meaning – Guess when we're thinking of shooting this scene?

Producer - "Will we have to ADR this?"

Meaning – even I can tell this is awful for sound. Oh the cost! Can you fix it with magic?

7

NO UNDO BUTTON

Planning, Problems and Pitfalls

"No plan survives contact with the enemy."
Helmuth von Moltke

You should try to plan ahead; you'd be a complete pillock otherwise, but I'd advise having a plan B stuffed in your back pocket as well. And some sweeties, but not squishy ones.

You could be in a boat that hits a rogue wave soaking the entire kit; in the middle of the Masai Mara and run out of fuel; your assistant might have a mental breakdown; you and your gear could be shat upon by a troop of baboons; you could be bitten by a poisonous snake; get totally lost in the mist at midnight on a small island with sheer 100m cliffs all around; find yourself trapped in a small plane where the undercarriage refuses to deploy; get stuck in a tiny lift up to the location with all the kit and a fearful old lady; or have car stunt go so spectacularly wrong that it nearly lands on your head – you can't plan for that sort of shit (but it happens nevertheless.)

Most Sound Recordists, like Boy Scouts, are notorious for being prepared. They have almost as many useful things in their pockets as Camera Assistants, though Camera Assistants always have more pockets. Always. No really – don't get into a competition. I suspect some Camera Assistants actually have Boy Scouts in their pockets. Like a handy snack.

If you are working on a drama you will usually be given the option of attending a recce. This gives you a chance to do some actual planning ahead, as opposed to complete guesswork.

See Appendix 3 - recce checklist

PP&P

Studio vs Real world

Everyone has their preferences, but I don't know a single recordist who doesn't like to get out of the studio every now and again, and
that's despite all the problems you have to face. A proper sound stage will be fully isolated against noise from the real world. Even cobbled-together sets in warehouses provide some attenuation. You may have a battle vs aircon, or electrical hum, lighting and acoustics but it is as nothing compared to the real world. Strangely, the quieter and more perfect a sound stage is the more you will hear the sparks farting or
the supporting artists chatting away on their mobiles behind the flats.

PP&P

Background noise - ongoing and transitional, is the enemy.
Sometimes you have a measure of control and sometimes you don't.

I once had to contact the head of a British location lighting company to get one of their 80Kv Generator vehicles fixed because it was vibrating so badly. Recordists had complained before for weeks via production, I was told, but unfortunately nothing got done.

That evening I emailed the guy directly (I happened to have known him as a spark from the distant past) and cc'd everyone concerned (production especially) likening the thing to an ice cream van droning away right next to the set and pointing out it was not fit for purpose. It had the desired effect. It wasn't just a complaint – it made the lighting company look unprofessional, and that Generator was off the shoot the next day and back to London to be fixed. The sparks actually thanked me - they had been chocking it up and doing anything they could to reduce the noise for weeks, and that is not their job.

Thanks Chris.

But I'm afraid it's not always as "easy" and ongoing background noise will range from the annoying to the downright impossible. Transitional unwanted sounds (such as aircraft) merely means suspending shooting until it has passed, but agricultural machinery, for example, clanking from one end of a giant field to the other will make your life very difficult indeed.

Much will depend upon the type of production you are working on (period drama, for instance, is rather unforgiving of, say, jet engines, strimmers, motorways and text alerts.)
You can take aerials and satellite dishes off cottages but you can't stop Heathrow.

You'd need something like a Pandemic for that.

Ah, the irony.

PP&P

Weather

Your main problems are wind, rain, sun and extremes of temperature. There are now some amazing products available to reduce wind noise: you just have to put your hand in your pocket. It's worth it though.

Rain is difficult, though there are products that help. It is tricky to filter the noise out because it is broad-spectrum. Although there are now clever tools available to reduce background [e.g. Cedar] I personally would never employ them on set except possibly in the monitoring chain. You cannot reasonably afford the time and effort needed and post production will do it better. You'd have to give them the non-treated tracks anyway. Rain is noisy, wet and annoying.
I hate it.

Water also gets into everything – even the best kit gets sodden in unremitting conditions. Sometimes it goes "phzzztt!" which is bad. You can hear that tiny crackle and know it's time to change out the whole cable run; the dropout that means the radio mics are on the absolute edge. You should know when the weather is likely to catch you out – rehearsing in cloud and shooting in full sun is not just a problem for the DOP, as every boom op will tell you. You want dry, cloudy, still.

A sorry situation for sound is when the camera and the artist/presenter are sheltered (by umbrellas, a canopy, etc.) and you can't see the rain because it's not backlit but you can sure hear it. A sensitive close-up exterior scene under a large umbrella will make you wonder what you have done to offend the filming gods and will probably end up being ADR'd (unless it fits the scene, which is unlikely.)

Precipitation will (eventually) kill your mics and soak into everything. Whether you are running radios or cables this is a major problem, and hoping it will go away is not an option – I am guilty of this. If it's the mixer then you are really unlucky because it should never have got that wet in the first place, but if you have been forced to stand in a downpour for twelve hours what can you do? You have to go through the whole chain to find out where the water got in and this takes a load of time. If at all possible replace everything and get the sodden kit somewhere it can dry out. If the 1st AD is looking at you and pointing at their watch... use your puppy eyes or something.

Cold – extreme cold needs very careful planning. Cables can snap, mic capsules freeze and batteries die within minutes. You honestly need to do some serious research if you are going to be recording in conditions worse than -10C

Heat – is not quite such a problem. You will most likely die before your equipment does.

I would love to hear from any sound guy who's gear expired from dry heat before they did (corollary: I do not want to hear from any sound guy who DID expire before his kit.)

Humidity, on the other hand, is potentially *extremely* harmful.

And wind doesn't just affect microphones. It can blow big, heavy things like sound carts off cliffs.

PP&P

Acoustics
Large echoey environments can sound great – but not for dialogue or a piece to camera where it will be difficult for the audience/customer to hear. For a single presenter on a radio mic expressing their wonder at the subterranean Roman reservoirs all is fine – in the end, production can always use subtitles. But you can't usually do that on a drama. Radio mics can help you, but not much. Long reverb times are a menace. Can the artists can lower their voices? Well, not in a shouty scene. Best of all - get heavy drapes erected.

Caves can be disconcertingly dry-sounding (it depends on the rock); forests can echo too, but worst of all are man-made structures (like stairwells, warehouses, and stations.)

The club bar at Woodnorton had a weird acoustic. Due to the hemispherical bowl which was part of the roof structure you could clearly hear conversations from across the other side of the room. Until a certain A-course turned it upside down. No idea who that was.

PP&P

Water

I've mentioned rain, but solid bodies of water hold their own problems. The health and safety issues I have addressed elsewhere but essentially it's pretty obvious that water and electricity do not mix – the water always wins. It's bloody unpredictable too. You'll cry if your cart is pulled overboard; especially if you are hanging on to it.

Luckily you are mostly dealing with low voltage DC but that might not always be the case. Should your equipment become submerged then 12V DC or 48V phantom will not endanger life but it will bugger your gear, probably terminally.

Most of the time it's not even your fault.

So there we were doing a scene in an upturned car in a ditch at the bottom of three fields. It was soggy and muddy but otherwise workable. It had rained the night before but the local landowner had advised location dept. that our situation was safe and that the trickle of water through the ditch was usual. We set up cameras and microphones, all in locked positions, strapped the artists in upside down and started filming.

All was well for an hour or so and then the trickle became a stream and that quickly became a flood. The water from the fields had taken hours to filter into the gullies.

Within two minutes the location was flooded. The 1st AD called it in time and thankfully artists and crew escaped safely – not so for the equipment.

PP&P

Ground surface and tracking shots

Lumpy fields present a problem for camera and boom operators, especially if muddy, or full of rocks or rabbit holes, furrows or crops, which, as you have already guessed, applies to most of them.

A long pre-laid track across such ground is always possible and you will have to plan accordingly, perhaps laying boards for both dolly and boom-op.

Hard ground and gravel are much easier surfaces for the crew but present you with the horrible problem of multiple crunchy footsteps over the tracking two-shot of a cosy conversation up the drive to the Manor House. It's not so bad if it's raining – one of those rare situations where the rain may hide your problems – but generally something to really watch out for. Carpet might be the answer, but it depends on the shot. Or get production to relay the whole drive with rubber gravel, (given the budget). They'll love that.

It's all in the planning but never forget that directors change their minds a lot. Or the owner of the location (who is being paid handsomely by the way) will come out and shout "Why is that fellow walking on my lawn? I thought we had an agreement! And no tip-toeing through the bloody tulips!"

PP&P

Another thing about locations - you need to respect them and the people who may live there.

At best, the unit may not be allowed back if you annoy the locals. Worse, you get stuff thrown at you. Or they turn up their sound systems and open the windows. At times like that, you feel glad you are not a location manager.

At the other end of the scale, the surroundings may be so posh and precious that the slightest scuff mark will cause all hell to break loose.

PP&P

Clothing

Sometimes you have input – but mostly not, because it's not about you is it? There are some materials that are simply inimical to sound (metallic, plastic, starched, liable to static, creaky and old leather, etc.) and it is worse for personal radio mics.

In a drama you simply have to go with what needs to be worn.

Squeaky shoes – afflict both artists and crew. Professional crews will usually be aware, as will Costume Designers, but often someone will arrive on set with lovely new trainers. Supporting artists are often the cause of unwanted clippy-cloppiness, or mysterious regular squeakiness. Get your crew to listen and to identify the culprits during rehearsal because it is nearly impossible to do so from your position because, even with the best of monitoring systems, the annoying sound is probably off-shot. Once again, an experienced crew will save you earache.

PP&P

Lighting

The problems very much depend on what kind of production you are working on.
Boom shadows – this is where the shadow of a boom microphone and/or the pole flits across an artist or the background or both when on shot.

It shouldn't happen most of the time given a professional crew. There are lots of ways of curing boom shadows caused by the lighting setup.

Quite often, it is the result of the sun emerging from a cloudy sky and then "bam" you're screwed. Hopefully, so is the DOP, because the scene will no longer cut: this will either involve relighting, construction of a huge silk, or a lot of waiting. Use this time wisely (by rigging radio mics.) OK this might not always be possible (outdoor pool scene for instance) so talk to the director and DOP instead – they want to get this shot even more than you, and may be willing to change the angles.

PP&P

Camera, dolly and boom noise

Thankfully modern cameras don't have sprockets or run tape any more, but there are a load of peripherals that are still noisy, especially focus servos which have an intense ring to them if not in pristine condition.

If you are expecting a lot of close, intimate work you are going to have to get on top of this, either by insisting on a change of camera focus kit or, more reasonably, asking if the focus puller can handle it manually (sometimes the old ways are best but it depends on circumstances.)

Dolly noise is usually (though not exclusively) confined to wheel or seat squeak. A good grip will solve this for you before you even mention it. A not-so-good grip will need to be nudged. Most dolly noise can be solved with WD40, but not always, as in the case of creaky floors in old buildings. In these cases it's down to ADR or replacing the dolly with a hot head, post production magic, or re-imagining the scene, whatever is within budget.

Handling noise – the bane of most recordists' lives this is the unwanted sound that travels up the boom pole and into the microphone. It is directly affected by the quality of the microphone mount, the boom pole itself, the mic leads and the skill of the boom operator.

In the vast majority of cases, it is the latter.

Skilful, quiet boom operators are worth their weight in computer chips but they already know that. It seems that no amount of expense on peripherals and new devices will cure a fumbly boom op... at least, that's what I have found.

If you can hear the trainee wobbling and shaking through your headphones then he or she needs a lot more well, training.

PP&P

Vehicles

How much more can we stuff in before it won't take off?

Not just cars, but trucks, trains, buses, boats, planes and helicopters– all can be lumped into this problem category. Not that there is any real problem; it just takes a *lot* more time to rig and de-rig things. I have always looked on it as a bit of a fun challenge. Every department will take longer to do their thing – especially cameras and lighting – so you have quite a bit of "what if" time, which means you have time to consider "what if this or that" happens and you plant a radio bug accordingly.

Later on, if it turns out invaluable, you feel really smug and vindicated. If not then so what? You wasted no-one's time.

PP&P

Animals

Mostly this has to do with health and safety but certain animals (especially horses and lions) can be skittish regarding boom microphones equipped with furry dogs. Many "filming" horses have been trained to be cool with loud noises, sudden moves and microphones on poles.

Most lions have not.

Neither have elephants. More dangerous are hippopotami, water buffalo, snakes, ticks and mosquitoes.

Oh and rabbits (more children were bitten by rabbits on The Really Wild Show than any other "dangerous" animal. Safest of all were the utterly docile piranha who refused, to a fish, to go into a feeding frenzy.)

Afternoon dump – by the elephants of course

PP&P

The general public

This is largely a Production problem and depends very much on where you are shooting. However, you always have a duty of safety.

Get used to: "What'ya filming mate?" or "Wos goin' on 'ere then?"

Your immediate thought might be - "Oh piss off what has it got to do with you anyway?" but NEVER say that out loud.

Security

Despite a security presence, do not assume you, your team, your equipment or your vehicle(s) are entirely safe. It's sad but it's a good idea to lock everything all the time, even in the most pleasant-seeming circumstances. Ensure your assistants always lock the vehicle(s). If there is no evidence of forced entry then the insurers could well argue against any claim, and lost gear could cost thousands to replace, let alone the headaches regarding the next day's shoot. There are bad people around who target drama film crews. Location Dept are obliged to leaflet the area beforehand giving shooting times and dates and, to some, this is like an invitation to hang around and see what opportunities arise.

One fortunate recordist had the whole contents of his van nicked. The thieves, having realised they'd get next-to-nothing down the local pub for this weird kit, decided to try and sell it all to a very well known sound equipment company in the same area. The company, highly suspicious, played along, and eventually the thieves were arrested, red-handed, and the recordist got his gear back. Lucky Richie. Well done Pinknoise.

PP&P

Equipment failure

This will inevitably happen, usually at the most inopportune moment, according to Sod's Law.

Possibly the worst failure is when you have been tempted to tinker with your gear on set.
DO NOT DO IT! You may well have an hour to spare where the director is rehearsing a particularly long and involved scene.

For God's sake, drink tea instead.

Do not start investigating the sub-sub menus of your digital recorder. Do not try and fix that niggling problem with the headphone distribution. Because it will all go horribly wrong, and you will spend the sweaty-est hour of your life trying to get back to square one.

Believe me, please.

Other failures are simply down to Fate, but as long as you are dealing with professional i.e. extremely expensive kit, it is rare for (say) a microphone to just "stop working" without warning. If it does, it has been dropped or knocked over by someone who has covered their tracks by propping it up in the same position and keeping quiet.

Keep calm and investigate the probable causes of the failure – a blown battery fuse; a broken solder joint; a bad set of dry cells. Hopefully this will be during rehearsal time. Digital problems are often "bitty"; analogue connection issues are often noisy/fizzy and intermittent. If all else fails you are going to have to explain to the 1st AD or the Director, Engineering Manager or whoever (depending on the type of shoot) that you have suffered a technical breakdown and that it will take "x" minutes to plug in your emergency spare.

You do have an emergency spare, don't you? Please say yes.

If that doesn't work you are going to have to replace everything in the audio chain one bit at a time. DO NOT let your crew get into headless chicken mode. You and they need to be methodical under extreme pressure.

Easy to say - much harder to accomplish.

Having said that about things rarely just stopping working I recently suffered a completely mysterious radio mic transmitter failure. One day fine, the next day dead. No sign of damage, no water involved. Nothing. Just fate. Or gremlins.

.

PP&P

Sense of Humour Failure

This can happen to anyone (quite often after equipment failure un-funnily enough) but I am talking about a general malaise here. It is usually a result of a succession of events, and a kind of shadow begins to loom over the shoot. There's a time to have fun, and there's a time to be serious, but there's never a time to be oppressed. Just try and recognise when it's you that has had the sense of humour failure in the way you treat others. I very much doubt it is you (the recordist) as the root cause because, frankly, you're not that crucial to morale. But then maybe you are? In a small crew, at night, miles from anywhere you could quote from "Life As A Sound Recordist" and cheer everyone up.

No matter what – it's just filming. That's all it is. It might be a documentary or a commercial or a drama or the Biggest Epic Ever but it is JUST recording stuff for entertainment at the end of the day.

Unless it's live. That's fun.

Even so, no matter how important it feels at the time, history will record it as an amusing gaff if things go wrong. In fact, the more it goes wrong, the more likely it will go viral. Look on that as a good or bad thing as you will.

I have done lots of live stuff.
It's just entertainment – it's not surgery.

*Admission time: I have worked with the likes of Stewart Morris and Michael Morris (no relation.) No matter what I said above, there are some people for whom the show **is** everything and is **not** just entertainment. I daresay you'll meet someone like this in your career at least once. They are forceful, frightening and unforgiving characters - at least on the outside. I worked with Stewart on a job that criss-crossed the country auditioning "talent" for the main show. It was fairly easy and involved playback and radio mics - but I decided at one point in the first week the main vocal mic needed a battery change though I was unfamiliar with the design (production provided all the kit.) I took it apart and broke it.*

So... I could run, kill myself, or admit the truth. Eventually I chose the last option and this guy with a monstrous reputation just rolled his eyes and got his gopher to get me another (a model I knew how to battery-change.) I thought it was the end of my career. He didn't care - it wasn't the main show. He cared more about the wine at dinner that evening.

I think, despite everything, he knew it was just bloody television in the end as well.

Michael Morris is a director/producer as intimidating and furious as a bear with toothache, but he gave me my first break as a drama sound recordist so what can I say - except a heartfelt thank you, Michael.

There are good shoots, and there are bad shoots, but a little planning goes a long way.

We were filming on an abandoned airfield early in the morning when I noticed my recordist was getting a bit twitchy and distracted. It wasn't a long scene, but the director was doing take after take. The longer it went on, the worse the recordist got; jiggling in his seat; looking agitated and uncomfortable.

At long last the director yelled "Cut! Happy with that."

The whole crew then watched in amazement as the recordist bolted for his estate car, and roared off down the runway. Unfortunately the tailgate wasn't shut properly and all the remaining sound equipment was dumped in a long line on the concrete.

I use the word "dumped" advisedly. There is nothing to hide behind on an airfield – not a tree, nor a bush.

Plan ahead – bring a cork.

8

SHORTCUTS

Tips & Tradecraft

You can have good, quick or cheap - pick any two (and other tips)

It's all about the money.

In the vast majority of productions there will be a limited budget. This means you won't always get what you want in terms of crew, equipment and pay. It's up to you to determine what you'll accept, and when someone is taking the piss. There are, of course, big budget projects around, and it's wonderful if you can get one of those gigs, but big money does not guarantee professional success. It is largely (though not entirely) about luck – better to work on a low-budget unexpected wild sensation than a multi-million dollar turkey in terms of your future career.

What would you rather put on your CV? (Yes I know you want to put "Bond", but anyone who has ever worked on the Bond or Star Wars franchises always manages to slip it into the conversation somehow.)

T&T

Try and learn a little about the basics of every other craft on set.
This is a bit of a tall order, I admit. I was given probably the best education by the BBC at the time and count myself extremely lucky because my initial audio course involved experience of film and video editing, TV studio craft (cameraman, floor manager, director, production assistant, racks operator) and time spent as an audio assistant in Bristol meant I got to know sparks, makeup, costume, riggers, camera assistants, and production runners and I made an effort to understand what their jobs involved.

You might be a natural – empathetic and understanding of other's problems but if you haven't the slightest idea of what they actually do or what their priorities are you are going to be in a poor position to argue your point in the event of a conflict of interests.

The more you know, the better position you will be in. Sound is not the most important craft in the hierarchy of film and TV production, and it's a good idea to remember that.
There is nothing to be lost and everything to be gained by asking other people on set about their jobs. There will be plenty of time to chat during rehearsals before the "crew show" so why not make it productive?

Chat to the security guys. They are often local and know the area - best bierkeller; cafe; worst place to park? Talk to make-up about the actors - who is lovely and who is not; which of them is a problem? Get friendly with the sparks - need a magic arm at short notice? There you go mate. Have an erudite discourse with the DOP about the esoteric word-of-the-day and oh by the way how is he proposing to light the night shoot?

The opportunities are endless and it is not just about extracting information you need; it's about understanding what other people do, what is important to them, and what sort of problems they have to deal with, and how you could, possibly, help.

T&T

If you are going away on a filming trip (it doesn't really matter where) pack a treat, and share it at the time of lowest moral. If someone on the crew has been particularly obnoxious, then you have a choice, don't you?

Out of the kindness of my heart I bought a director a box of Liquorice Allsorts^{tm} because he had alluded to them and he was a really nice guy. Later that day he said, if I let him do two very different shots together at the same time, I could have an allsort, I protested: "You can't bribe me with my own gift!". Which he took as assent. Love ya David.

T&T

Check the import regulations if working abroad, and be careful of flight restrictions regarding tools, certain batteries and aerosol cleaning products. Security is only going to get tighter, so check before you get to the airport. Good idea to check if you have a mic as well.
(Hi Marc H)

T&T

Overlaps
This is where one actor's dialogue overlaps another in a single shot. There is no inherent problem with this as long as you are covering both (or all) actors, but it can and does cause problems for the editor.

I spent years editing sound so I do have a clue as to what can and can't be done but if the director wants the actors to overlap (and they often do for genuine artistic and performance reasons) then it's almost impossible to ensure they overlap at exactly the same points on multiple shots.

My answer – don't worry about it – it's the director's decision, unless they say "should we do that again without overlaps?" to which the answer is, always, "yes".

It is easier for everybody that overlapping dialogue can be recreated in the edit suite, but sometimes there is no artistic substitute for allowing the actors to do their thing, driven by the moment and their own craft and sense of timing.

T&T

Wildtracks – every recordist knows the value of these non-sync recordings. I would recommend making a list of what you need during the shooting of a scene and put them in order of importance from "crucial" down to "would be nice".

Nobody else on set apart from your crew will quite understand why you have a particular wildtrack under "crucial" and why you are so insistent upon it. Mostly they don't understand what you do at all. (Ah ha, the answer - tell them to read this!)

If you have the time then by all means explain to the 1st AD, director or even better, the producer, because what you are doing is saving time and money. Usually, they won't want to know. Unless you are working on a big-budget movie where 99% of the soundtrack will be delivered in post, even so, a lot of the time location sound and performances can be unique and there is only one chance to capture it. Or a second chance – with a wildtrack.

Certainly most things can be synthesised but it is

easier for everyone and definitely more satisfying to be able to hand over the real unexpurgated actual lion-fart recording.

T&T
Lions

Mostly they are sleepy and don't give a shit. However, get slightly too close and dangle a furry object over them and they wake up really really quickly.

It's then you realise you are in an open-topped vehicle with no protection whatsoever with a boomful of tasty-looking bait in your hands ten metres from a killer.

Seriously though, there is an attitude common amongst sound and camera and production persons that can develop over time that somehow your job makes you immune to the realities of the situation – be it war, natural disaster or wildlife.

I just slowly and gingerly put the boom away and lived to record another day.

T&T

Critters

Beware meerkats – they are intelligent, inquisitive and want to steal your shoelaces.

T&T - The tip that Never Was

The List

This is a physical or virtual record of all the people you have worked with that you deem important to your career. Ideally you want a name, project, their job description, contact details and your private assessment of them. How you do this is up to you but make sure it is totally private or you may be breaking database rules.

The List will help you assess who might be important to contact in the future regarding work, and who you might want to avoid. If you store this electronically then security is paramount – it must be password and virus protected.

If you write it in a notebook – the Data Protection Act 2018 still applies as far as the notebook must be kept in a "lockable" filing system. But this is all best practise anyway. A good rule of thumb might be – your List doesn't exist. The very last thing you would want is to share it.

So you don't have one, *do you*?

Yup, it's all kept in your head, which is the best place.

9

WARNING, ARE YOU SURE?

Health and Safety

Close to the edge?

Boring? Gets in the way of efficiency? A secondary consideration? Program Prevention? Think again, because these days you can go to jail, and certainly be sued, if you don't take your responsibilities seriously regarding H&S.

"Can't I just be a recordist?", I hear you say. Not really, I'm afraid.

The very act of filming can be hazardous. Because you are all concentrating so hard on your jobs it is all too easy to be unaware of potential dangers, but you have a legal responsibility to ensure all risks are minimised – it's not all down to the producer, though they often have to take final responsibility.

Are the people in the picture safe? They have life-jackets, but what is the drop?

As a sound recordist you might be head of a team – and that means you are responsible for their safety too. You have to try and imagine what witless thing your boom operator might do before they do it, otherwise the finger of blame will be pointed at you, and not at the idiotic boom op, because he or she was presumably doing what YOU asked them to, last anyone heard before they disappeared with an "Oh bugg..!"

"Why?" I hear you say, "Surely that's not fair? I can't possibly think of everything."

Doesn't matter in the eyes of the law. It's bloody difficult, but not actually impossible, because the law does *not* require you to imagine the insanely unlikely, but to consider reasonable possibilities.

In case you were worried, of course everyone in the picture is wearing a life jacket and is tethered except, strangely, the stunt co-ordinator. But then, they have a lifetime of special training behind them.

The unseen drop is a very slippery 45 degree 10m rock face into a cold sea.

The Risk Assessment

On any production you should be given the opportunity to make a risk assessment before shooting. If not, you will have to assess the situation as it evolves. This can be one of the trickiest of circumstances you will face.

> Eg.1 On a drama recce you note that the scene in question requires an actor to run across a field shouting at some far off miscreant (say: "Gerrof my land you ill-mannered layabout..." etc.)
> How will it be shot? Will your boom operator have to run backwards in a field? Are there rabbit holes or roots or small boulders? What if it is wet when filming? If you do not have a plan and a risk assessment in place, and your boom operator sprains his ankle it could be you that is taken to court if it is found you didn't even consider it a problem. The actor is NOT your concern.
>
> Eg.2 There are just the three of you – camera, sound and director in an unusual jungle situation. The light is fading fast, and you are about to exceed your daily hours. You are willing to ignore the extra time but there are wild animals, snakes and insects around. But the Night Monkey is proving elusive. If you say you'd rather get back to safety and call it a

day, will you ever be employed by that company again? You feel unsafe, but you don't want to be thought of as a wimp. But why are there just three of you? Where is the local guide? He should have been part of the team from the start of the day, but you didn't say anything.
And what will the cameraman and director think of you? You are supposed to be a team. They think there is a once-in-a-lifetime shot to be had in the next half hour, but the sound too, could be unique (and only you will have captured it.) How do you assess the risk?

Eg.3 The film features a chase involving a very particular, rare and expensive car driving at extreme speeds. Unfortunately it is only available for a few days but the last one, thanks to production, is a "sound" day.
It's relatively easy, from a sound point of view, to rig the car with recorders and microphones – but what about the risks? It's highly likely you are not the primary recordist – you have been brought in specially for a day to do these important wildtracks. What are the rules of the road in this country? What are the rules regarding private racetracks? Can you justify your place in that car for a once-in-a-lifetime opportunity? Come on – you know you're tempted. It's all do-able.

Risks are not always huge and life-threatening, but there's the problem. If you do not take reasonable steps to mitigate them you really could end up in court, even though you discounted them. It is, to be frank, a minefield.

Risk Assessment – Hazard: requirement to work in minefield – Mitigation: sound coverage will be by remote microphones (any possible equipment damage to be covered by Production) – sound team will be located in bunker 0.5 miles from set with radio link.

I once said in a RA involving a helicopter shot that sound would be safely concentrated around the tea urn but that didn't go down particularly well with the H&S executive (though it earned a few laughs). What was wrong with that? I told them where we would be and that we would be safe, but humour and H&S mix as well as oil and water.

These days I nearly always include a warning about bitey insects. Just in case you dismiss bitey things - mosquitoes alone kill more than a million people a year. Remember that if you are filming in a malarial arial (Doris BBC canteen).

At least you can voice a concern these days and be taken seriously and not worry about being seen as a whistle-blower and compromising your freelance career. That is a *major* advance.

Health and Safety - Children
Working with children is now a very specialised and controlled occurrence (as it should be, given all that has happened).

The procedures in place are for their safety and your protection. Specific to sound, young actors may, for example, need to be fitted with radio mics. It is not my place to explain every procedure here, just to point out that this is a very important area which you should and must learn in order to protect yourself and your team, and ensure the safety and comfort of children on set. Liaising with the specified chaperone is crucial.

This is serious stuff and non-compliance can end up with a prison term.

Cleaning and Sanitising Equipment
The need for such procedures has never been so great and will remain important even in a post-Covid world because everyone has a right to expect that the equipment you supply (especially personal microphones and headphone monitoring) is both clean and safe to use each and every day.

You should institute procedures that means your gear is sanitary. Isopropyl alcohol and de-mineralised water solution is very useful and effective but be aware it can damage microphones, foam windshields and cables.

UV-C light sources will kill viruses and bacteria but must be used carefully to avoid harmful radiation affecting the user or your precious mic capsules. You must see that your assistant(s) are trained in the safe use of such equipment or you (as HOD) will be held responsible

Are your lavalier mics IP58 rated?

Do you know what that means?

When it all goes Horribly Wrong

No matter what, you cannot anticipate everything - if you *can*, you should be anointed a prophet, and make a lot of money prognosticating rather than messing about recording.

The cameraman, DOP, me and the stuntman/pilot (in dress and wig) were shooting in a light aircraft.

We finished the dramatic in-flight scene and decided all was good and we could head home to the little-used airfield and show the director the rushes.

After a while we noticed our pilot cranking away at something between his legs. Minutes passed (in silence) whilst we circled and then he admitted he couldn't get the undercarriage down. The irony was not lost on us – the drama was about an aircraft crash. The hydraulic system had failed and the manual over-ride was jammed.

He hadn't wanted to alarm us, but explained he needed some sort of metal tool to try and free the undercarriage. Our plucky DOP started to dismantle his monitoring gear whilst the cameraman and I looked at each other and nodded in unspoken agreement that if we were going down, we were going down filming, just in case the tape survived.

The DOP, after a lot of unscrewing, (with my screwdriver) somehow produced a long piece of metal tubing that the pilot used to jimmie the undercarriage into position manually and we landed without mishap.

It does make you think though. There wasn't an awful lot of fuel left either.

Why was I up there? Because the Director disliked flying and there was a spare space – and I could save post a lot of trouble with some sync internal aircraft acrobatics fx.

Could the problem have been anticipated? I don't think so – planes are checked thoroughly for obvious reasons. But equipment fails all the time.

This time I must admit it was a bit inconvenient.

Did anyone panic?

No, we were all strangely calm – I cannot explain why, except that you just don't do panic in front of your colleagues.

In fact, maybe I'm weird, but the prospect of a wheel-less landing on a runway surrounded by ambulances and fire appliances was strangely exciting.

But thank you Graham, for your quick thinking, because crashing into a foamy airstrip with no wheels probably wouldn't have been any fun at all.

10

MASTERING

Ups and Downs

Life on the road as part of a film crew – to many people, this is what it is all about – travel, excitement, challenge, new experiences, new friends, and a sense of achievement especially co-operatively against adverse conditions – it can be tremendously rewarding. There are life-changing opportunities to be had and you get to enjoy it and, extraordinarily, to be paid for it. This is the job at its peak and best and it does happen though it is never *quite* as glamorous as people believe.

The fruits of your labour can be sweet indeed but don't totally buy into the hype and the hope. At its best this job means you will be with people you like and trust and who enjoy working with you in a challenging environment, yet where you can socialise after a rewarding day's work in a place you would never have contemplated visiting – maybe never even have heard of – and perhaps you'll get the chance to relax together in the evening, and enjoy the surroundings and the local cuisine.

On the West Scottish island of Islay, in the late afternoon I found myself wandering along a sheep track across the gorse and heather from one remote Time Team site to another. It was chilly, damp and misty but I was wearing the right gear and wasn't cold or wet. I stopped and looked around at the deserted moorland and suddenly felt completely... content. I have never forgotten that moment – nor the whiskey, the Céilidh, and the finest scallops in the world. But that moment alone was exceptional.
(Regards to Steve S)

The downside – this is the crap situation that drives you nuts, and about which you have little control.

Physical discomfort; the realisation that the team you are with are not the people you thought they were; the breakdown of a sense of humour, or reality, decency or humanity,
and communications.

It undermines morale. You wish you were anywhere else. Throw in a few technical problems and you have the recipe for a nightmare.

The only answer is to soldier through and add certain names to the "bad" side of your non-existent list. Or leave. But you won't be back.

I was having an egg sandwich in the back of an open-topped Landrover parked under some shady trees in Kenya when a whole troop of baboons above decided, maliciously, to empty their bowels upon us. The shit-shower missed the cameraman and producer in the front but covered me and my equipment.

Baboon crap is very sticky and smelly. Luckily (or unluckily) it didn't hit the recorder. However, we still had six hours or so of filming to do before sundown which was punctuated, every thirty minutes or so, by the cameraman or the producer going:

> "What is that awful smell?" to which the other hilariously answered:
> "I believe it's the recordist in the back".

Ha bloody ha.

I spent a long time combing out the dried-up plop from my 816 furry dog, but when we finally arrived back at our camp I had a near transcendental experience –
although the showers were simply buckets of warm, wood-smoked water that you operated with a string – I cannot describe to you how wonderful it is to wash out a headful of baboon-poo and then go and have a bottle of cold beer (even if it did have some kind of horrible dead worm in the bottom. No one else had a worm!)

I really didn't give a shit though.

Happy days.

Life on the road will also put a strain on relationships with loved-ones.

Why?

Because you (clapper loader, runner, recordist, operator, director, producer, whoever) are fulfilling your career dreams; enjoying a great experience; networking and meeting new and attractive people; learning interesting stuff, stretching yourselves and you are away from all that ordinary boring domesticity.

And for considerable amounts of time and (hopefully) for considerable amounts of money. If this sounds like a world of life-changing opportunity then of course it does. It is very appealing, and you do feel guilty. Sometimes. Not all the time.

Filming can and does break relationships. It doesn't have to – but it often does, because your other half is left at home alone for long periods, possibly lonely, bored, resentful and even envious. I cannot offer any remedy – just be aware. Location filming can be addictive. Sometimes it is almost impossible to turn down despite whatever else is happening in your life.

The problem is – if you are in the Marines and they say "you are going to serve 3 months in Belize" and you have a week-old baby, you have little choice. You say to your partner -

"Sorry but I've been ordered to Belize for three months. I'll see you and babes every day online luvu etc."

If you are a freelance director, recordist, cameraman, or AP say, then the choice is a total shit-bastard.

"Errrm... darling I know we've just had a baby and it will be tricky but there's a three month job in Belize I can't turn down if we're to pay the mortgage and stuff. I'll get in touch... if I can."

IDCOL

There is a saying regarding location filming - "It Doesn't Count On Location"

This is bullshit.

There are, of course, numerous opportunities for romance. Shagging, to make myself plain. That's fine as long as you remember
that nothing goes un-noticed within a tight film crew.

If you really want to do the costume assistant or the make-up team of course you can (as long as it is consensual) but please don't for a minute believe it's going to stay secret. Let's face it, there are a lot of staggeringly attractive people in this industry. OK I know it sounds un-pc to say so but it is true - it's intrinsically tied-up in the very nature of the business.

You don't see it down the local supermarket.

And it is true that marriages/relationships within the industry are fraught with problems too because you may love your partner, but you also love your job, and it has just offered you the opportunity to work in sunny Guadeloupe for four months, but your partner's booking has just fallen through.

What do you do?

Is the relationship more important than the money? These dilemmas happen in all walks of life, yet it seems particularly prevalent in this industry because of the allure of exotic locations and, worse perhaps, exotic people.

There doesn't seem to be any kind of job barrier between relationships either - though there may be a well-hammered bed and a mirrored ceiling, there appear to be no inter-departmental limits - Props Master and Costume Designer; Camera Operator and Clapper Loader; Boom Operator and Third Assistant Director; Grip and make-up assistant; Director and dresser; even the bloody caterers get involved.

As for the sparks – well, you *really* do not want to know.

I suppose it's the freedom. And stress relief. Yes, that's what it probably is.
Where did I hear that before?

Don't say I didn't warn you is all I can advise in a worldly-wise and avuncular way. Heh. And you thought filming was just about the job... ha ha ha - it's WAY more than that.

You get to be part of a team and the closer the team the better the result, but at the end of the shoot there is the breakup. Maybe you'll all be re-united – but who knows?

Lows

If I think about it carefully the lowest point for me in my career was not on a desolate location or a boring studio or in a smelly B&B, but at home, wondering what I had done wrong to have had such a bad year. Worse yet, everyone was about to have the Millennial celebration of a lifetime.

Everyone can have a crisis of confidence in themselves, their abilities, and the future. Actually, you can see no future. And you simply can't magic up work from nowhere, no matter how "positively" you try to think about it.

Not every freelancer has work perfectly lined up for the rest of the year, despite how well-known you are. The fact that you cannot be in two places at once is also *really* inconvenient. It's just not much talked about - these dry spells, because every freelancer has to maintain a facade of success. Admitting to not being busy is like confessing you have some weird and possibly contagious career-disease.

Even if you have managed to fill your diary, the world then just goes crazy and throws a pandemic at you. Yes, everyone suffered, but freelancers in the Film&TV Industry were hit more than most. Shit happens. But a wise cameraman (Sandy) quoted something to me – one door closes, another opens, and he was right.

If you are still interested, against all the obstacles

Worming your way in

I am sorry that I can't give you the magic formula for success. If I could, I would be a very rich and happy ex-sound recordist writing my third book in my second home in the Bahamas, where I have most enjoyed filming (more than Wales, certainly.)

There are University courses, there are a few traineeships. The usual progression is trainee sound assistant, sound maintenance technician, boom operator, sound recordist. If you have read this guide and thoroughly understood everything in it, maybe you can start straight out from scratch as a recordist? I wouldn't recommend it, but it might be possible. And above sound recordist I suppose you could aim for directing, though it is an unusual route.

At the time of writing, probably the best courses (in the UK at least) are run by the National Film and Television School. There may be other, equally worthy contenders, and this is only my personal opinion having worked with numerous aspiring trainees. You are not guaranteed a ticket into the industry but will learn aspects of the craft that are actually useful, as opposed to theory, or "film studies" or "The History of Cinematography" which are all very well, but don't teach you where the REC button is.

However, it must be said that my favourite characteristics in a trainee are not a know-it-all attitude, but common sense, enthusiasm and intelligence.

As a much-missed colleague of mine, Dave Baumber, used to say - "Some of them are about as much use as a chocolate teapot."

If you don't run, don't ask, and look like a rabbit in the headlights when confronted with the unusual - you won't last. Be quiet, attentive, on-the-ball and don't tell me what I already know. Demonstrate your capabilities - don't boast about or exaggerate them.

And please don't disappear with the van keys even if you are having a personal crisis. Give them to me and go home to your distraught lover.

No one is irreplaceable.

Realities of freelancing

If you can get a PAYE job in sound then go for it! These kind of jobs usually refer to sound maintenance or boom-opping but the advantages are huge – you get a regular salary and statutory rights (sick pay and holiday pay) and a chance to build up your experience. The downside might be that you are stuck on one programme or project for a long time and miss out on networking.

Freelancing is both financially and mentally challenging! I wrote this guide during the Coronavirus Pandemic of 2020/1. Apart from a bit of gardening and some equipment maintenance there was nothing else left to do. The industry shut down, and none of us in film & TV drama earned a penny for months.

My company made its first loss in 25 years.

That was an unusual period, but you need to plan for disaster even during normality. There is no holiday pay, there is no sick pay.
Not only do you not get paid if you don't turn up, the word will quickly spread that you are unreliable.

The work available is also unreliable, and if you suffer from a cancellation, it is often a fight to get anything from the production company that have screwed you over. And you will never be compensated for the job you turned down in order to be free for the one that disappeared.

You don't know when the phone will ring and you never know when or if you can book a holiday.

If you enjoy skiing, think of the consequences of breaking a limb, and read the insurance fine print very very carefully because it hardly ever covers loss of freelance earnings.
Together with "Ups and Downs" it might look like "Realities of Freelancing" is trying to put you off – not so. I am simply being as blunt and honest as I can.

When things are going well then it can be great, but it can also be depressing and incredibly stressful. Even the money you make in a good year isn't all yours - remember the taxman? A year later you will still owe that money.

At the height of the Coronavirus pandemic HMRC would agree to defer your tax bill, but they never write it off.

The job can also affect massive life decisions, like moving house or having a family. If you never plan to do either then great, but they do say "never say never". If this subsection sounds a little too melodramatic then I can only say I have been through it and know an awful lot of others who have as well so discount this advice at your peril. I know you will if you are a twenty-something.

Next thing you'll be telling me you haven't thought about any kind of personal pension!

Hearing your stuff broadcast

It is only on very rare occasions that what you have recorded will be broadcast straight to the world. Usually it will have gone through a post production process. When you hear "your" sound it will probably be substantially different to the original. This is because it needs to conform.

Just be content that something you have contributed to has been broadcast to thousands, possibly millions.

As long as they can hear the words, what's wrong? Well, a lot if what you recorded was an orchestra, but I hope you get my meaning. There is a lot of gratification to be had from hearing your sound on the TV or cinema/theatre.

In a way, it's there for ever, and yet it's ephemeral, so bask in it whilst you can.

Networking and maintaining your profile

It need hardly be said that maintaining a profile is as important part of freelancing as it ever was – it's just that times and methods have changed, and will continue to do so.

Once you have inveigled yourself into the industry, you need to put some work in to maintain your image and particular mystique (if you have one). However, all the clever web-sites in the world will not compensate for a good session in the pub with the producers or a curry with the crew.

Although I was unaware of it at the time, I was told later that the best thing I ever did for my career with a particular long-running drama was organise the occasional (and unbelievably popular) charity pub-quiz for the entire cast and crew. That was networking in one of its finest forms i.e. nobody realised. Social media is important, but never discount personal discourse because that is what separates you from the rest who are all trying the same thing – they are knocking at the door. Very loudly.

But you are already in the building.

I am too much of a dinosaur to presume to teach anyone about social networking. If you read this guide you'll find a lot of stuff in here that you won't get elsewhere. Use it to your advantage as much as you can.

Don't be afraid to ask intelligent questions.

As a last resort, read the instruction manual.

The author in a reasonably good mood even on a boat, it would seem

Getting out

At the end of the day, when it all gets too much, you might need a bailout plan. Some recordists go on to run their own gear-hire companies. Or buy a camping site. Or a care home. Or a fast car.

Then there's always writing, gardening, or mending broken kit.

But really this book is about the beginning and not the end.

Avoiding a Negative Profile

It is advisable to never -

- Assume you know best
- Openly criticise someone in front of their peers/the crew
- Accept that a situation is safe if your senses are screaming otherwise
- Rely totally on one single piece of equipment
- Let a director bribe you with your own Liquorice Allsorts^{tm}

Living The Dream

You are, as a person, having read this far:

Enthusiastic, Reliable
Resiliant, Competent
Friendly
 … and now so much more knowledgable.

If I were a little bit more clever, I could distil "sound recordist personality" more perfectly.

Maybe rigorous and technically competent would creep in there, but I don't want to make out that "sound recording" is some special thing – it isn't, in itself. Lots of professions need these qualities. Paramedic, airline pilot, waitress and hairdresser come to mind.

So what is it that made me want to do this strange job, and why might you?

Well, I love buttons and faders. I really do. I suspect it's a childhood thing.

I like the fact that even though I do the "same" job each time it is always, unfailingly different and the challenges never stop coming.

I would like to say I like recording nice sound, but most of my life is spent listening for sounds I don't want to hear. It's different I suppose if you are a Nat Hist specialist and the mating call of the African Black-Bellied Bustard is your thing, but then there's probably a long-nosed frog spoiling it.

It is rather wonderful to listen to voices though - not the ones in your head, but the ones in your headphones.

Just hearing Jenny Seagrove, Sharon Gless, Kwame Kwei-Armah, Christopher Eccelstone, Tom Hiddleston, Martin Freeman, Ray Winstone, Sorcha Cusack or Simon MacCorkindale was a privilege. I would have loved to have recorded Sean Connery, Oliver Reed, Fenella Fielding or Richard Burton... if only!

You get the chance to travel and meet the most extraordinary people.

You also get to leave a legacy of work that just might survive you.

At the time of writing, the Covid-19 panic has taught me that the entertainment industry has a very important role to play in society; it's not *just* bread and circuses, it's about mental well-being. We still don't know what the fallout will entail, and may not for years. A lot of productions have found clever (if expensive) ways around the problems although drama in particular has been affected disproportionately and for obvious reasons.

We are not saving lives – just facilitating entertainment.

Yet though I am sometimes inclined to the cynical, I think that drama is an essential expression of culture and as such, there will always be -

Lights! Camera! Sound! Action!... and a fulfilling and challenging career to be made in all.

Oh and don't forget the caterers - they are very important.

Ah, hot puddings.

YAWN.

Has he bloody finished yet?

Really?

Can I fart now please?

No, I'm too bored to roar.

Is that furry thing on your pole edible?

It looks edible.

It even smells of...

Is it baboon?

APPENDIX 1

Being a short boring list of essential sound bits most of you will want to skip over.

>XLR M-F short
>XLR M – M
>XLR F – F
>XLR Y cord M – 2F
>XLR F – 1/4" mono
>XLR M – 1/4" mono
>XLR M – RCA
>XLR F – RCA
>XLR – open ends
>XLR3 – miniXLR (TA3)
>6 pin XLRF – 2x XLRM
>5 pin XLRF – 2x XLRM
>3.5mmM – 3.5mm F extension
>RCA F – 1/4" mono
>XLR M – BNC
>rca M – M
>rca F – F
>rca M – 1/4" F
>rca M – F
>lemo – BNC (TC out)
>lemo – BNC (TC in)
>BNC M-M (barrel)

As for the future see:

https://www.canford.co.uk/Audio-and-data-connectors

Beware *connecting balanced to unbalanced cables. Beware sticking 48V phantom up the arse of something that may object or, possibly, fail spectacularly. Beware 4-pin connectors as they are usually 12V DC. You will see there is no 3 pin to 4 pin connector in the list. Beware RCA/phono and 1/4" leads as they are prone to picking up every stray signal around. Beware BNC connectors - they seem to break just by looking at them. Lemo connectors are insanely difficult to wire and are consequently expensive but you still need spares. Beware modern CAT5 connectors (and similar) - the slightest bit of dirt or damp will render them useless. Beware creating earth loops. Ensure your mains equipment is PAT compliant. As for the good old XLR - they take a beating, and will need replacing from time to time. Beware splits in cable casings - this will allow water ingress - gaffer is only a temporary solution and self-amalgamating tape is better. Or you should split the cable and add new connectors. Warning - everything breaks - everything. And it will be seen as your fault; your problem because (usually) you are supplying the gear. Be prepared and have spares. Lots of spares.*

Answer to the 3.5mm problem - use a suitable small diameter self-tapping screw about 15mm long, and your sound screwdriver.

APPENDIX 2

Glossary – nowhere near comprehensive, of course, and I can only vouch for the English versions of film-speak. It might be quite fun to learn this list and then test people who pretend to know what they are doing, though that would be a bit evil.

> 10-1 – the person in question (usually an actor) has gone for a piss
>
> 246 – a wooden block with three depths for adjusting the height of props (in inches)
>
> ADR – automated dialogue replacement in sound post production
>
> AP – assistant producer – fancy name for a terrible job
>
> Barn doors – physical barriers to control a lamp's throw or area of effect
>
> Boom – a very expensive extendable pole on which is hung a microphone
>
> BNC – ubiquitous though unreliable video connector originally designed for the military
>
> Buzz – non-sync room atmosphere recorded to aid post
>
> Buzzy – unwanted soft focus
>
> Chinese – to make the barn doors horizontal; to use the sync from one take over another (mostly obsolete)

Choke – noisy sparks equipment required to run most lamps – luckily becoming redundant as technology marches forward

Cocktail Hour – the time at which very senior and well-respected Directors like to finish – usually about five o'clock

DAT – Digital audio tape – a flash-in-the-pan defunct recording format

dB – decibel, a measure of relative sound volume

De-rig – to tear apart and store as quickly as possible without actually breaking anything

DFI – we have had a bit of a rethink and have come up with a new and surprising way of proceeding that nobody expected (stands for different fucking idea)

Dolly – heavyweight wheeled camera platform

DOP – Director of Photography; a very important person indeed

Drone – flying camera platform so noisy that sound department gratefully go home early

EQ – stands for equalisation and generally means messing with the frequency envelope

Ext – Exterior

Extra – supporting artist that fills in the background with interesting over-acting

F – female form of a connector e.g. a socket

Fill – soft light source that literally helps fill the set with illumination

Flat – section of the set supported by props and stage weights

foley – sound effects made by a person in a studio added in post production

French flag – a device to cut down unwanted light spill

Furry dog – windproof cover for a microphone

Gaffer – head spark; heavy duty tape

Gash – rubbish (usually recording tape)

Gobo – lighting effect utilising a physical light mask

Grip – person in charge of moving the camera around on shot, usually on a dolly

HOD – head of department

IEC – electrical connection standard, commonly mains

Int - Interior

Jib – camera arm and/or camera move

KBS – the usual situation at the end of the day, it supposedly stands for "kick bollock scramble"

Key – a lamp that it is important not to stand in front of, especially if you are a boom operator.

Lollipop – member of Properties Dept.

Low loader – a specialised vehicle for travelling shots

M – male form of a connector e.g. a plug

Magliner – a big, heavy, expensive trolley-thing for camera dept. but no one knows what the name means

MOS - mute; without sound. Or mutual oral sex

NAB – North American Broadcast standard quarter inch reel size adopted almost worldwide

Nat Hist – Natural History, often wildlife, but including the whole of the natural world

OB – Outside Broadcast

Overrun – to exceed the allotted shooting hours

Pag – a small wooden support structure often used by Grips; short for Paganini

Pan – horizontal plane camera move

Post – Post-production – the process of assembling, editing, dubbing and grading after shooting

Prop – almost anything on set that isn't technical and is supposed to be seen on camera.

PTC – piece to camera (e.g. solo news broadcaster)

Quarter inch tape – an old system of sound recording on reel-to-reel machines

RA – risk assessment

RCA – aka phono : domestic audio/video connector now largely defunct

RF – radio frequency

Rx - receiver

recce – a preliminary reconnaissance of locations prior to shooting.

Shoot – the whole rather important business between pre-production and post production.

Silk – large translucent structure used by the DOP to reduce contrast. Quite fun to watch the sparks try and control what is basically a huge sail on a windy day.

Spanish – after the not-so-famous Iberian archer "El Bow" this means – get rid of.

Spark – electrician and eating/drinking/farting machine

spl – sound pressure level or, as I prefer, sound panic level

SSD – solid state device. Lightweight,

reliable – a dream come true. Yeah right.

Steadicam – a specialist camera rig worn by an operator (rather like the M56 smart gun in Aliens)

Strawberry filter – pretend filming

Sync – short for synchronous sound where dialogue and effects are in time with the picture

TC – timecode

Tilt – vertical plane camera move

Tx – transmitter or transmission

Village - a set of covered video monitors about which the Director, Script Supervisor and a group of HOD's and hangers-on collect all trying to see what is going on

WA – wide angle camera shot

W/T – wildtrack: non-sync sound recording that will help post production

Wrap – wind reels and print (supposedly) in other words time to de-rig and go home

XLR – a connector used extensively for professional sound (eXternal Line Return)

ZOOM – a variable focal length telephoto lens or a proprietary name for videoconferencing

APPENDIX 3 - *A Recce Checklist (lite version)*

A recce consists of a load of people standing about in some forsaken location talking about things that may or may not happen. In this instance (an exotic beach with a warm, inviting ocean and a wonky horizon) we used somewhere else completely different on the day, which kinda buggered my plans entirely. Ho hum, get used to it. Everybody had to adapt. It's all part of the fun.

Things to Look Out For

Int noise – fridges, chillers, aircon, buzzy lights, boilers, mysterious hums, underground trains – controllable? Where are the sockets? How long can freezers be unpowered? Liaise with location manager and local owner.

Int floor (creaks?) carpets? Hard floors. Does it matter? Eg only if dialogue

Low ceilings, narrow corridors, tiny rooms, mirrors – other boom impediments?

Is shooting area really echoey? Eg stairwell, large hall, arena. Shots – do we know, as an audience, where we are?

Have design built a tiny set in the middle of a massive warehouse... damping required?

Ext noise – traffic, construction sites, public, piped music, nearby schools, trains, fountains, waves, volcanoes – what are the sources? How bad? Can they be controlled? Do they need to be controlled (ref to script – eg is dialogue intimate or shouted?)

Public roads – will there be traffic control?

Pavements – unit cannot impede public access. Is the road flat, or on a hill, and how busy will it be at time of shooting? What is the general noise like – even, uneven, occasional, intrusive?

Cables in public areas – r/mics required? How much production control envisaged?

Aircraft flightpath – wind direction?; more than one aircraft per two minutes? (no one else will notice this, guaranteed)

Where will video village be? Identify dead areas (i.e. those the camera will never see) Can cables be used safely in this area? Where will the genni be parked on each location – does it really need to be so close? Liaise with gaffer.

Moving vehicles – low loader, A-frame, cam on board? How can we cover this sound-wise correctly and safely? Will vehicle shots be static "rock-and-roll" (greenscreen/monitors?) Or moving? Windows up or down? What speed? Open top? Engine type? Rare vehicle? (will need wildtracks.)

Int /ext location floor surfaces – is it safe? Is there a trip hazard to a boom operator walking or

running (perhaps backwards)?
Or anyone else on the team (bricks, shit and used needles are not uncommon)
Other – Steadicam, drone shots, minicams, quad bikes, speed boats, jibs, wirecam, helicopters, int. planes, etc.?
Safety: trip hazards, slopes, water, heights, overhead electrical wiring, special fx, explosions smoke, traffic, stunt vehicles, animals on set, disease, insects, rodents, extreme weather, underlying medical conditions, ground conditions; mitigation of all of the above - how can it be achieved within budget?
General – it is almost unheard-of for a location, once decided, to be changed because of sound issues so just make notes and don't stress unless filming under a waterfall or in a minefield or a bottle-making factory (or The Royal Mint - loudest background noise I ever heard in my life.)

My least favourite locations:-
airports, train stations, bus stations, pedestrian bridges over dual carriageways, warehouses, meat processing plants, piers, amusement arcades, funfairs, zoos, sewers, classy chic shops, grungy chip shops, motorway services, shopping malls, small boats, swimming pools, schools, restaurant kitchens, bottling plants, recycling facilities, construction sites, military training grounds, auctions, stairwells, raves, caves, naves and anywhere near waves.

Thanks Bestest Crew for many fun, happy, amazing years.

Thanks to Daffyd Llewellyn

Thanks also to AndyMac & Vic with whom I share some thoughts and a bottle of wine from time to time; Richard North who took me all over the place and didn't care how wet I got; "Wiz" Wilson for his enthusiasm and encouragement, Mike Smythe for his mentoring; Sophie, Kirsty, Abdul, James and Ramon for their awesome boom skills; Alan for his Audio Unit leadership and to much-loved work and drinking companions Bob Lasseter, Sandy Tristrem, Clive Lovell, Geoff (Ikegami-El-E-Mac) and the BLU crew. It has also been a pleasure to work with Jaz Castleton, Graham Jaggers, Simon Walton, Andy Parkinson, David Innes Edwards, Jordan Hogg, Trevor Gosling, Rod Lewis and even some of the sparks. Oh and thank you Helena who still puts up with it all.

All images copyright Tony Briskham unless otherwise credited.
All cartoons copyright Alix Briskham unless otherwise credited.

An ERRASTRUNE Production

Absolutely no lions were harmed in the production of this book.

They slept through the whole thing.

'Tis rumoured there was once a map to **The Lands of Errastrune**.

It is long lost.

However, fragments of tales persist and rumours remain.

Before their destruction, the Fabled Lands encompassed the Log Sea, the Focal plain, and the Land of Abdul, where the people walk around with their hands held above their heads in a puzzling ritual.

Overlooked by Camera Mount, somewhere was hidden in a grass valley the Video Village - a build upon the banks of the Edit Channel, reached by a clapper bridge. A plethora of craft would bob there in the bay under the quay lights, and they would often venture out at high tide into the black, level waters of Roll Sound, propelled by silken sails; the crews with fishpoles, reels and nets, hoping to nab a bite, and not sink.
From the Village, adventurers looking for action would follow the Honey Wagon each morning as it went rolling up the wild track to Younit Base. There, it was said, the Ktring tribe would offer any who wanted to line-up in the queue their fill of breakfast, with pan-fried chips and peppered Spanish sausage, if you could choke it down. Passing by the Fixitin Post you would find yourself up the sides of Dolly Creak on the way to the scenic flats. Or, as some hotheads might, you could keep turning, shoot past the Depthof field, and perhaps make-up time by taking the short director's cut, through the barn doors to the stream of rushes, providing it wasn't blocked.

At this stage you certainly needed to wrap up extra well and keep a grip on the guide track for the route was error strewn and even if you reached the budget allotments you needed to take care as there were scattered around a lot of holes in each plot.

Eventually, with stiff legs, some cheap biscuits, and a determined character you would reach your desired location – marked by a sharp sandy beach populated by scampering monitor lizards and crabs, with a soft background buzz of crickets, the drone of countless insects, stunted trees and lines of weed coated in film cast upon the strand.

But you had made it for the climax of the day, despite the hardships. Everything was set...

and against the flare of the light bouncing and sparkling on the sound waves, you might just filter out (if you had timed it correctly, and were given the right direction)... an outline on the horizon.. a frame... a flicker-free vision... in that golden hour...

> A volcano.
> A huge boom.
> A looming shadow.
> Oh bugger me!
> It's actually gone off.
> That's not in the script!!

RUNNING...   AT SPEED.

and...

CUT

Printed in Great Britain
by Amazon